WESTERLUND 2 ● NASA, ESA, THE HUBBLE HERITAGE TEAM (STSCI/AURA), A. NOTA (ESA/STSCI), AND THE WESTERLUND 2 SCIENCE TEAM

SPACE IS COOL AS FUCK

SPACE IS COOL AS FUCK

Kate Howells & Friends

Andrews McMeel
PUBLISHING®

Space Is Cool as Fuck text copyright © 2017, 2020 Pantera Press. Created by Martin Green. This edition © 2020 Andrews McMeel Publishing. All rights reserved. Printed in China. No part of this book may be used or reproduced in any manner whatsoever without written permission except in the case or reprints in the context of reviews.

Andrews McMeel Publishing
a division of Andrews McMeel Universal
1130 Walnut Street, Kansas City, Missouri 64106

www.andrewsmcmeel.com

First published in Australia in 2017 by Pantera Press Pty Limited; this revised edition published in 2020.
P.O. Box 1989, Neutral Bay, NSW, Australia 2089

20 21 22 23 24 TEN 10 9 8 7 6 5 4 3 2 1

ISBN: 978-1-5248-6297-8

Library of Congress Control Number: 2020942024

Author: Kate Howells with contributions from Garrett Johnson, Laura Pumphrey, Madeline Glowicki, and Scott Koenig
Editor: Kevin Kotur
Production Editor: Elizabeth A. Garcia
Production Manager: Tamara Haus

ATTENTION: SCHOOLS AND BUSINESSES
Andrews McMeel books are available at quantity discounts with bulk purchase for educational, business, or sales promotional use. For information, please e-mail the Andrews McMeel Publishing Special Sales Department: specialsales@amuniversal.com.

Contents

Introduction	1
The Big Bang	2
Exoplanets	6
Speed of Light	10
Energy	14
Matter	18
Black Holes	20
Multiverse	26
Atmospheres	30
Gravity	34
Galaxy Smash	38
Antimatter	42
Dark Matter & Dark Energy	50
Aliens	56
Radio Waves	60
Bill Nye Interview	64
How Rockets Work	74
Space Stations	78
Just How Big Stuff Really Is	82
The Moon	86
The Apollo Turd Transcript	90
Tycho Brahe	96
The Mars Rovers	100
The Voyager Missions	104
Living Off-Planet	110
Microorganisms	120
Earth's Magnetic Field	124
The Sun	128
Venus	132
Saturn	136
Big & Small Space Rocks	140
Pluto	146
Jupiter	150
Moons of Jupiter	154
Heat Death or Big Freeze	160
Index of Artists	167
About the Author	170

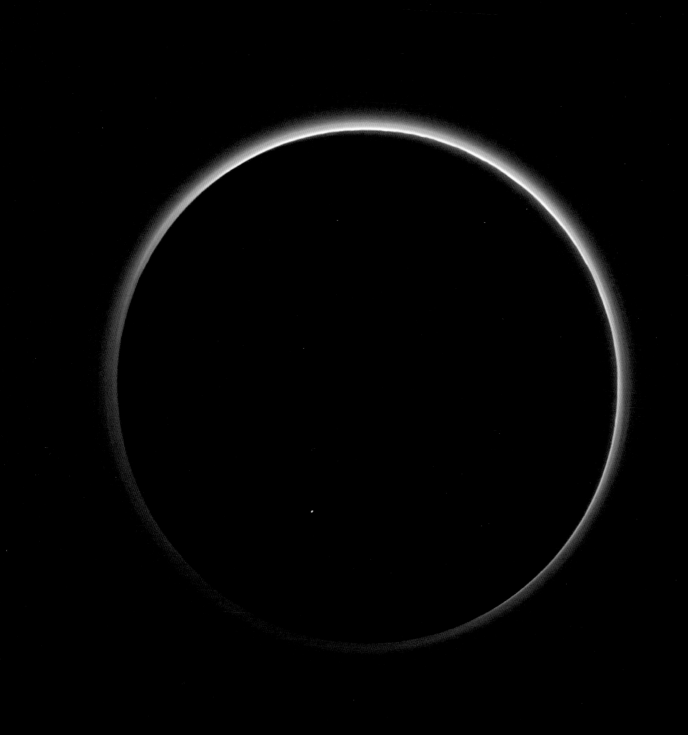

BLUE HAZE OVER PLUTO ● NASA/JHUAPL/SWRI

Introduction

Space is fucking dope. The universe is big, wonderful, and crazy. It's huge—incomprehensibly so. You and all your friends and everyone you've ever loved and all the cool shit you like, you're in it, this one gigantic thing that just exploded out one day and started being. And it's got volcanoes, aliens and dinosaurs, and rocks hurtling around at mind-boggling speeds, flying through the cosmos, each their own little world. Some are completely inhospitable, some have so little gravity that you can basically fly, and some have diamond storms. Anything that you can even imagine is happening somewhere up in this bitch right now.

This is no textbook. We love space, and like fools in love, we want to shout it from the rooftops. We want to share our love with everyone, but science has a tendency to be a little tough to digest. Which is what this book is for. We are not experts but simply enthusiasts—we want to give you a little taste of the glorious reality you inhabit by providing an introduction to some of the incredible stuff out there, told plainly in the tongue of the common people.

We hope this will encourage you to learn more. The people of this planet are connected by a magical network of information systems that we call the internet, where you can dive right in. The deeper you dig, the cooler it gets.

Learn some shit, and tell your friends. 'Cause this space stuff is out of control.

One of the most pressing questions we've ever faced as a species is where the fuck did this all come from? How did this mind-blowingly huge collection of matter and radiation that we call the universe—from galaxies that are a million light-years across down to the quarks inside of an atom—come to be? According to smart people in white coats, it was a rapid, spontaneous expansion of space called . . .

THE BIG

BANG

OK, hear them out.

In 1929, Edwin Hubble (the guy the big space telescope is named after) observed that faraway galaxies are moving away from us. Not only that, but the farther away they are, the faster away they move. He took this to mean that all observable parts of the universe are moving away from each other. Galaxies are still moving away from each other today, so that should mean that they were closer together and hotter in the past. Scientists proposed that this expansion could be traced back to a single point, from which a cosmological clusterfuck erupted in every direction. Hence, the "big bang theory."

People were skeptical, naturally. When someone tells you that we're flying on a giant rock through a continuously expanding space bubble of matter and radiation that started as a subatomic singularity 13.8 billion years ago, you're likely to scoff. But almost all of the evidence that physicists have gathered supports the idea—same with cosmic background radiation and Einstein's theory of general relativity.

So what the fuck actually happened during the big bang? Here's our best guess.

First, there was a singularity. Singularities are points in space and time that can't be described by our laws of physics, like black holes. They're currently a big red question mark on physicists' drawing boards. But anyway, from this singularity we all of a sudden get a hot, dense, rapidly expanding chunk of universe. And we mean rapidly.

Our universe started out as an infinitesimally small bowl of soup, so to speak, where gluons and quarks flew around and smashed into each other. This soup was so hot and compact that matter and energy were practically the same thing.

Then, as it grew in size, our soup settled ever so slightly, and quarks became neutrons and protons, the building blocks of atoms as we know them. And as all this was happening, our little bowl of soup ballooned in size from the atomic level to 100 billion kilometers across.

Oh yeah, remember when we said this was rapid? All we just described happened in less than a second.

Hydrogen, our first element, soon formed. The universe at this point was about 10 billion°C, which is over 600 times hotter than the goddamn surface of the Sun.

Things slowed down a bit from there. In the next few minutes, the universe got cooler and less active. There were no stars yet, and all this hydrogen gas prevented any visible light—though it didn't matter because there wouldn't be anything with eyes for well over 12 billion years.

Then, over the next 300,000 years, hydrogen gas clumped together, gravity squeezed the fuck out of it, and we got our lovely stars and galaxies. Hydrogen gas dissolved into a plasma that allowed visible light, and we basically got the universe as we know it.

Ta-da!

CHECK THIS SHIT OUT

The universe has continued expanding ever since the big bang got things going. It's getting bigger and bigger all the time, and it's actually speeding up. Which is a pretty wacky thing to think about. The universe is, by definition, everything that exists, including empty space. So if it's expanding, what's it expanding into? What was out there beyond the edge of the universe before the universe got big enough to occupy that space? We don't have any answers for you!

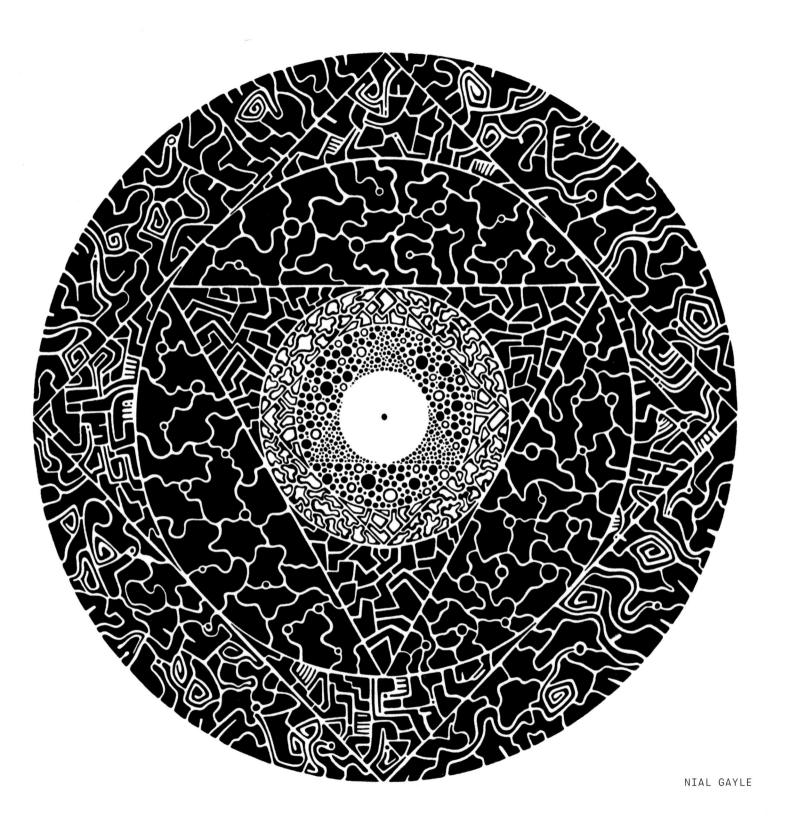

NIAL GAYLE

KOI-961 EXOPLANET SYSTEM ○ NASA/JPL-CALTECH

HOLY SHIT THERE ARE A LOT OF *EXOPLANETS*

BRETT RANDALL

CAROLINE LEVASSEUR

Scientists have long known that the Sun is a star and the Earth is just one planet orbiting it. We know that there are shitloads of other stars too, but scientists have never known for sure whether there were other planets in orbit around them.

It sort of figures that there would be, but science likes to know for sure. Even our main man Carl Sagan dared only to speculate in the 1980s that there could possibly be other planets around other suns, but he went right ahead and imagined how spectacular they might be.

Eventually we got good enough at looking at space that we started finding them. In 1991, the first exoplanet—a planet around another star—was 100% for sure detected. A few more discoveries trickled in over the years, and in 2009, shit really kicked off. NASA launched the Kepler Space Telescope, whose sole purpose was to look for exoplanets. And boy did she ever find them. At the time of writing, over four thousand exoplanets have been discovered. We're finding that pretty much any star we look at seems to have at least one planet around it.

And that is a pretty insane thing when you think about it. That means there are trillions and trillions of planets in the universe. Exoplanet researchers have also found that about one in five stars of similar size and power to our Sun has a planet the size of the Earth orbiting at a distance that would sustain liquid water. They call this the Goldilocks zone because it's not too hot, not too cold, and astronomers are little kids at heart. This means that one in five Sun-like stars has a planet that could conceivably host life like ours. And in the Milky Way galaxy alone, there should be tens of billions of potentially habitable Earth-sized planets.

All of this obviously brings the question of alien life into better focus. If there are trillions of planets in the universe and life emerged on this one, the chances are good that life has also emerged on some others.

The unfortunate thing about all these exciting discoveries is that we're not going to be able to visit any of them anytime soon. The closest confirmed exoplanet is 4.25 light-years from our solar system. That means that if we somehow managed to travel at the speed of light, it would still take us more than four years to get there. And we don't travel nearly that fast. The farthest we've managed to send anything so far is the Voyager spacecraft, which took thirty-five years just to get to the outer edge of the solar system.

What we can do, though, is send radio signals in the direction of exoplanets that we think look cool. Signals go at the speed of light, so at least we have a chance in hell at communicating with anything that might be there. It's a slim chance that there's life on any given planet, and slimmer still that any life there would be able to detect and understand radio signals. But fuck it. We might as well try.

```
        CHECK THIS SHIT OUT

   The Milky Way has between 100 and 400
      billion stars. The Andromeda galaxy has
       1 trillion stars. NASA uses the word
     "zillion" to estimate the number of stars
       in space. No, we didn't make up that
    word: it literally means any uncountable
        figure. Australian National University
      gave counting stars a shot, estimating
     70 sextillion in space. That's 70,000
              million million million.
```

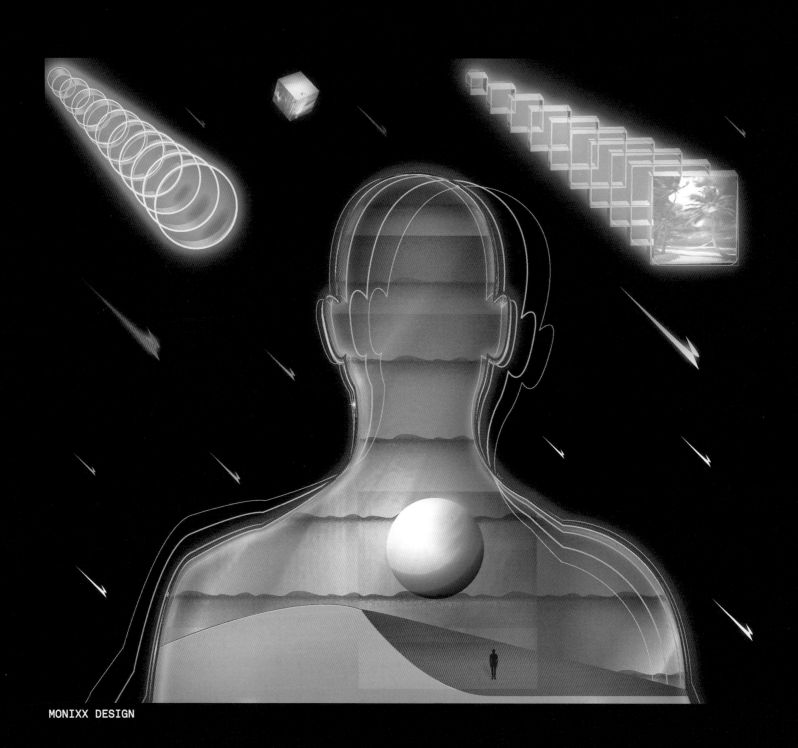

SPEED OF LIGHT

The speed of light in a vacuum is 299,792,458 meters per second.
This is the maximum speed limit of the universe as we know it.
Nothing can go faster than that. To us little humans jogging
around at about 2 meters per second, that seems pretty fucking fast.
Well, it is and it isn't. It's all relative, baby.

Light only takes 1.3 seconds to travel from Earth to the Moon. The fastest thing humans have ever launched from Earth still took 9 hours to get as far as the Moon. But if you had a buddy on the Moon right now, you could communicate with them with only a 1.3-second delay because information can travel at the speed of light.

So light is fast as hell, but the universe is not a small place. It takes light 8.3 minutes to get to Earth from the Sun, but that's still small potatoes. Light moving from our solar system to the closest star takes 4.24 years. Because this is such a stupidly large distance—more than 40 trillion kilometers—we just say that it's 4.24 light-years away. And from there, shit just gets wild.

Our own Milky Way galaxy is 100,000 light-years from one end to the other. The nearest neighboring galaxy, the Andromeda galaxy, is 2.5 million light-years away. So fuck talking to a buddy over there. That's one hell of a delay. The observable universe—what we can see from Earth—is a sphere with a diameter of 93.2 billion light-years. Think about this: the fastest thing in the universe, which goes at a speed incomprehensible to the human mind, would still take almost a hundred billion years to cross the universe.

And it has some pretty nutty implications. Because the universe is so enormous and information can't travel faster than light, the chances of contacting any intelligent beings that may be out there are pretty slim. Even if we spot signals of intelligent life coming from some distant solar system, those signals would probably have been sent out hundreds of thousands, if not millions or billions, of years ago. And vice versa. If you sent out a little "hello" to the universe and any intelligent alien ever picked it up, you would be long dead by the time they got it.

And forget about ever trying to go say hello in person. Say we found signals from an intelligent alien life-form at the very next star over, Proxima Centauri. We know that's only a few light-years away, so no big deal, right? We could just hop on over at the speed of light. Well, physics is a bit of a bitch, so no. Turns out nothing with any mass can travel at the speed of light.

So if you want to meet aliens someday, you'd better cross your fingers that we discover wormholes or some other way around the cosmic speed limit. Or hope that the aliens do, and that they're chill.

GAS REFLECTED IN VISIBLE LIGHT ● NASA/ESA/J. MAÍZ APELLÁNIZ (INSTITUTO DE ASTROFÍSICA DE ANDALUCÍA, SPAIN)

ENERGY

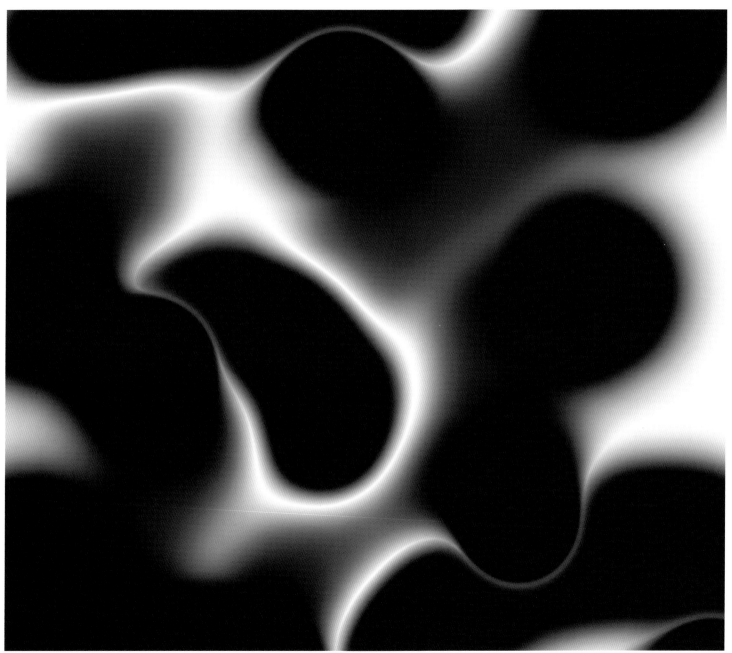

JAMES MARSHALL

A deep dive with an ordinary dude

**with guest writer
Garrett Johnson**

Lots of people talk about how there must be more to the universe than what we can see and touch and feel. There must be some other realm that we can only tap into in our subconscious minds or through psychics. There must be ghosts and spirits, auras and angels.

Well, if you're looking for another realm beyond ordinary matter, you don't need to get all spooky and crazy. As it turns out, there's plenty of cool (and real) shit beyond what you can see and feel.

Here to talk about the flip side of the material world is Garrett Johnson. Garrett isn't a physicist—he's an ordinary dude. He likes to stay up late drinking beers with friends, and he also likes to read about science. So here he is again to teach you all about what he loves.

What is energy, anyway? The thing your weird aunt summons when she's rubbing a crystal? The enlightened state you're vibrating at? The feeling of a mall cafeteria when you've eaten a bunch of mushrooms?

"Energetic states," "fields of energy," and "vibrations" are actually terms that describe completely real phenomena. But unfortunately, there is no evidence to suggest that the energy coming from someone's aura is in any way related to actual energy in the physical universe.

In science, energy has a fairly precise meaning, and it takes on many different forms. For example, heat energy is how much the particles of a thing are moving and vibrating and banging into each other; chemical energy is what holds molecules together; and nuclear rest energy is the amount of energy that could potentially be released from matter if it were to react with antimatter.

All of these things are sort of abstract, but they all relate to an ability to effect direct change in the physical world. And this is really what energy does: it changes from one form to another and in doing so makes a change in the physical world.

Think about a campfire, for example. By burning wood, you're creating a chemical reaction that releases energy stored in the chemical bonds of the tree created during its life. The bonds are fueled by sunlight—so when you burn wood, you're literally releasing stored light. The sunlight itself came from the fusing together of hydrogen atoms that were created in the big bang to form helium, in the process releasing some nuclear rest energy as radiation (light).

The story of the universe is a story of energy and the ways in which it changes itself to make things happen. Without these exchanges, absolutely nothing would ever happen, at all, ever.

Even matter, the stuff we can see and feel, can be put in terms of energy exchange. Everybody's heard of the insanely famous equation $E=mc^2$, but most people don't really know what it means. It basically says that energy "E" and matter (which is the only stuff with mass "M") are like two sides of the same coin and can be turned into one another.

Energy itself has no mass, but when it's contained within something (say by heating it up or making it move), it adds mass to the object. And likewise, when you take it away (by radiating light or slowing something down), it becomes less massive.

So when you weigh something that's hot it weighs more than when it's cold, or if you weigh something that's moving (and has kinetic energy) it weighs more than when it's still. In other words, when you're ripping down the LA freeway in your convertible soaking up the Cali sun on your way to a sweet beach party, you are actually slightly more massive than when you're standing still in a cold shower later that night trying to sober up for the drive home. The difference in weight is too tiny an amount to notice (maybe the mass of a proton), but the change is there.

On a much bigger scale, the Sun is constantly losing huge amounts of mass by all the light it's blasting out—around 4 million metric tons per second. Over its entire lifetime, so far it's lost around the equivalent of the mass of Earth, just through light radiation.

So more energy gives a thing more mass. And on the flip side, more mass gives a thing more energy in the form of nuclear rest energy. The energy contained within a thing is its mass multiplied by the square of the speed of light—a lot!

If somebody were to convert a 90-kilogram (200-pound) human body into pure energy (which is totally possible by reacting it with equal parts antimatter, where matter and antimatter cancel each other out and turn into pure energy), it would create an explosion 266 million times more powerful than the bomb dropped on Hiroshima. It's also theoretically possible to create matter from extremely large amounts of energy. Like taking those 266 million atomic bombs and condensing them into a person or a cloud of helium. Humans haven't quite figured out how to do any of this yet.

The deeper philosophical implications of this idea mean that matter itself can be thought of as extremely dense packages of energy.

Matter and energy are the same thing, and we live in a universe made solely of the same stuff expressing itself in different ways.

PUT THAT IN YOUR PIPE AND SMOKE IT.

MATTER

WHAT THE FUCK IS ALL THIS SHIT?

If your eyes could see the basic essence of all this shit around us, we'd be staring at the wall in astonishment all day. When we get down to the atomic level, this universe of ours gets pretty fucking bizarre.

Everything—every single thing contained in the physical universe—is made up of atoms, which are these incomprehensibly tiny things with a nucleus in the middle and a bunch of little negatively charged bits called electrons flying around the outside. The number of electrons flying around, as well as the number of neutrons and positively charged protons inside that nucleus, determine exactly what kind of element we have. Hydrogen is the smallest, for example, and it has just one proton.

And that doesn't apply just to all this shit around us.

It's us too! The average person is made of mostly hydrogen, oxygen, and carbon atoms: about seven octillion (or seven billion billion billion) in total. That's not a typo. Atoms are really little bastards, and it takes special equipment like electron microscopes to see them.

Even though they're small, they're quite spacious inside. If an atom's nucleus were the size of a marble, the farthest reaches of its orbiting electrons would be around 100 meters (over 320 feet) away. In fact, when we think of how we interact with the matter around us, the farthest reaches of electron orbitals are about as good as it gets. When we hold a rock, for example, our atoms don't touch its atoms. The outer shell of its electrons actually repels the outer shell of our electrons because they're both negatively charged. We're all just walking around in our own little bubbles of repellent electron shells.

So anyway, if matter is just a clusterfuck of atoms, how do we get all the different types of stuff in the universe, like planets and ice cream and fire and the air we breathe? Energy, that's how.

A bunch of atoms in their lowest energy state will form a rigid, interconnected structure—that is, a solid.

If we give them energy, say, by heating them up, those connections break down a little bit. The structure of connected atoms will loosen and start to take the shape of its container—voilà, a liquid.

Heat them up even more and those atoms will break loose from the structure and start flying around madly—boom, gas.

But wait, there's more.

Pump them with a shitload of even more heat and the electrons on these atoms will spill out of their orbits, thus making a soup of electrons that can conduct electricity and create magnetic fields. This soup is called a plasma, and we serve it every time we start a fire.

EVERY SINGLE THING CONTAINED IN THE PHYSICAL UNIVERSE IS MADE UP OF ATOMS.

BLACK HOLES

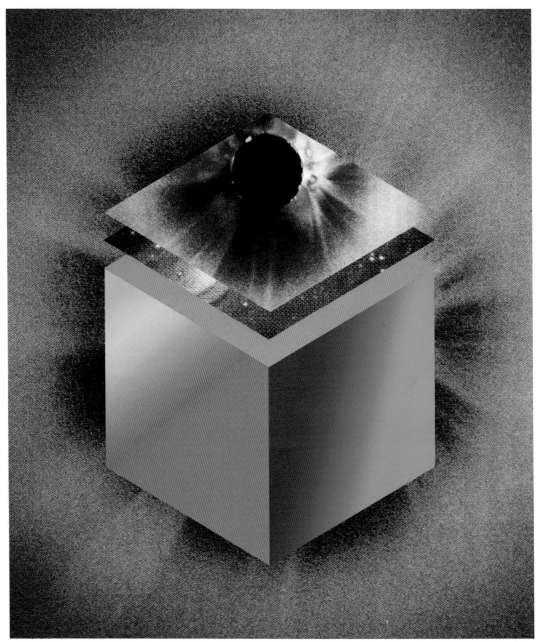

CHRISSIE ABBOTT

BRITNEY. LINDSAY. SHIA. AMANDA. MACAULAY CULKIN. THE LITTLE DICKHEAD FROM *TERMINATOR 2*. THE GOOD ONE FROM THE JACKSON FIVE (RIP).

SOMETIMES SHIT JUST GETS TOO HEAVY TO HANDLE. BIG MEGASTARS OF HOLLYWOOD OFTEN COLLAPSE UNDER THE PRESSURE OF THEIR OWN FAME, AND BIG MEGASTARS IN OUTER SPACE ARE NO DIFFERENT.

CHECK THIS SHIT OUT

275 MILLION STARS ARE BORN AND DIE EVERY DAY.

Every star is basically a nuclear reactor, using the gravity of its hugeness to slam hydrogen atoms together to form bigger atoms. This creates energy that makes the star shine and also balances out the force of gravity. You see, gravity wants to pull every part of the star in toward its center. But energy coming from the center pushes back out, keeping a balance. Every star has a limited amount of hydrogen in it, though, and in every star, this fuel will eventually run out.

In cases where the star isn't very big, like our Sun, this isn't a massive deal. The star just goes cold and chills like that in space until the end of time. But with bigger stars, once they lose that fuel, they've got too much mass to manage and it all collapses in toward the center. Sometimes this causes a supernova, which is basically just an explosion caused by too much star matter trying to be in the same place at one time.

But in supermassive stars (we're talking about 20 times bigger than the Sun), the collapse gets so intense that not even an explosion can counteract it. Matter condenses to the point that its gravity dominates every other force imaginable. There is no escaping its gravitational pull. Even light can't escape, and light doesn't even have any fucking mass! This, dear reader, is a black hole.

These things are cool, scary, and awesomely real. We can't see them (since light can't shine out of them or reflect off them), but we can see how they distort the space around them. We're even pretty sure there's an enormous one in the center of every galaxy. The one in the center of ours is 4 million times the mass of the Sun, but some are a billion times the mass of the Sun! They're pretty damn mysterious, and they've opened up some big questions in physics.

CONCEPT ARTIST'S RENDERING OF A CYLINDRICAL SPACE COLONY
NASA/AMES RESEARCH CENTER

MIKE MAKATRON

FOR AS LONG AS HUMANITY HAS EXISTED, WE'VE TORTURED OURSELVES WITH HYPOTHETICAL SITUATIONS.

From our perspective, life goes in a straight, fixed line. But we wonder about all the stuff that didn't happen: could things have turned out differently? We also think about how the seemingly tiny decisions we make each day may have a massive influence on the future. It turns out there are some physics behind this kind of wistful daydreaming. And it's a doozy.

You see, there may be a plane of existence alongside this one in which you didn't quit your violin lessons at age eight and instead went on to become the greatest musician in human history. There might also be a plane of existence where humanity has telekinetic powers or where cats take pictures of us and put them on the internet. Literally any goddamn possibility you can imagine—plus the ones you can't—could very well be occurring in a different universe at this very moment.

It's called multiverse theory, and it's pretty fucking neat. Here's the basic idea.

The observable universe has been expanding since it first came into existence around 13.8 billion years ago. If something is expanding, then it's finite and it should have a border, right? Then what's beyond the border of our universe? Cats using Instagram? A vast society of telekinetic superhumans? A happier and more successful version of yourself? Nobody knows. But some of the most prominent astrophysicists in the world have come up with some rad ideas.

We could be living in a massive slab of universes where each one rubs right up against the next in a grid, extending into the infinity of spacetime.

We could be part of a system of sequential daughter universes where every situation that has more than one possible outcome actually splits off new universes in which each outcome occurs. So every bad decision you've made has been made right in another universe!

No regrets!

Or maybe we're inside just one of an endless assortment of bubbles, each inflating at their own pace, each with their own physical laws and constraints. You hear that? If the physical laws and constraints we know so well don't apply to most other universes, it means the math and science you learned and have since forgotten may be useless garbage after all.

These are all just proposals, though. In fact, the multiverse theory has caught a lot of flack because many say it's scientifically untestable. How could we find evidence of a thing if we're locked in a grid or trapped in a bubble and can't access said thing?

Still, the idea has the support of some of the biggest names in space science, like the late Stephen Hawking and Neil deGrasse Tyson. So if you want to give yourself a quick pick-me-up by believing that there is a place somewhere, somehow, where you are a telekinetic violin virtuoso constantly being recorded by social media–savvy cats, then you have that right.

ATMOSPHERES

The atmosphere might sound like a pretty boring topic, and you probably had to learn about it in really boring ways at school. Remember its chemical composition? No, too boring. Remember what the difference between the troposphere and the stratosphere is? No, get the fuck out of here with that boring bullshit.

But the atmosphere is, in fact, cool as fuck.

We live in a sea of gas. We think of the air as just nothingness, and when we see car exhaust or steam coming out of a factory or whatever we think, "Oh, there're some chemicals pouring into the air." But no! Well, yes. But also, no! A car dumping its exhaust into the air is just mixing its gases with the gases that are already there. We're constantly surrounded by gas. Gas surrounds us right this instant.

There's actually about 480 kilometers (298 miles) of gassy atmosphere above us. That much gas weighs about a kilogram per square centimeter (or over fourteen pounds per inch), and all that is sitting on top of our heads. Conveniently, our bodies evolved to deal very well with this weight on our heads, and we'd actually have problems living without it. But you'd never think it was there, would you?

Most people think of planet Earth as the rocky thing with oceans, and think of the atmosphere as something that's just on top of that. The real planet is the hard stuff. But gas planets like Jupiter and Saturn are basically all atmosphere. There's no real difference between the outer layer of a gas planet and the atmosphere of a rocky planet. We're just living inside another layer of our planet.

The atmosphere is cool too, because it blurs the boundary between our planet and outer space. If we head out from Earth's surface to outer space, there's no single moment when we have decidedly left the atmosphere and entered the void. The atmosphere just sort of peters out, rather than ending at one particular place. And because the atmosphere is just the outer layer of the whole planet, it means that the planet itself doesn't have a clear boundary that divides it from space. We just decided that Earth ends and space begins at some semi-random point above us.

And the division between the Earth and its atmosphere is hazy too. The gases that make up the atmosphere are continuously being cycled through the Earth's oceans, rivers, lakes, groundwater, and all the life-forms on its surface. Every plant and animal takes in atmospheric gases and farts them out again in one deliciously stinky circle of life.

So not only are we swimming around in a gassy layer of our planet. We're also full of it, creating it, and making up part of a continuum from planet to space.

CHECK THIS SHIT OUT

With no atmosphere, Mercury's temperature fluctuates according to the Sun. It's as high as 425°C (797°F) during the day but drops to -180°C (-292°F) at night.

MOLECULAR CLOUDS MUCH DENSER THAN EARTH'S ATMOSPHERE ● NASA/ESA/N. SMITH (UNIVERSITY OF CALIFORNIA, BERKELEY) ET AL.; THE HUBBLE HERITAGE TEAM (STSCI/AURA)

RIA MCILWRAITH

CLOSE-UP OF A COMET TAKEN BY THE ROSETTA
ESA/R-SETTA/NAVCAM, CC BY-SA IGO 3.0

STUFF IS ATTRACTED TO STUFF.

THAT'S THE GIST OF IT.

When something has mass—that is, when it has a physical presence—it distorts spacetime around it; and other things with mass fall into that distortion, sticking to the massive thing. That is quite a simplification, but hey, we could write an entire book about how gravity is cool as fuck. The bigger something is, the bigger the distortion it creates and the more likely it becomes that other stuff will fall onto it.

We're used to this on Earth, to which we happily stick. But there are some cool examples of gravity doing things that we don't normally see.

Take, for example, this sick close-up picture of a comet. It was taken by the Rosetta spacecraft from about 10 kilometers (6.2 miles) above Comet 67P, and in the photo you can see boulders on its surface. The cool thing here is that there is no single "down" force of gravity acting on every boulder. It's subtle, but the various boulders you can see in the photo look like they're defying gravity, sticking to the comet at different angles.

This is possible because a comet is a very different place from Earth. Earth is enormous and it's rotating, a combo that causes gravity to crush all Earth's mass into a sphere and pull everything on the surface toward the center of the sphere.

But this comet isn't that bulky and hasn't crushed itself into a sphere like a planet or a moon. It has two big chunks sort of stuck together by a bridge. Because there are two big bulks here, there are gravitational pulls toward the center of each. So boulders on the surface can stick to one side or the other. It's a weird gravity party where we literally don't know which way is up.

GALAXY SMASH

THE GALAXY NEXT DOOR, ANDROMEDA, IS 2.5 MILLION LIGHT-YEARS FROM US, CONTAINS ABOUT A TRILLION STARS, AND IS EVER SO SLOWLY MOVING TOWARD OUR OWN GALAXY, THE MILKY WAY. (WELL, IT'S MOVING TOWARD US AT ABOUT 110 KILOMETERS PER SECOND—OR JUST OVER 68 MILES PER SECOND—WHICH SOME MIGHT CALL FAST. BUT IN SPACE, THAT'S SLOW AS FUCK.) IN ABOUT 4 BILLION YEARS, THESE TWO GALAXIES ARE GOING TO COLLIDE, AND IT'S GOING TO BE PRETTY FUCKING COOL.

When you think of two galaxies smashing into each other, you might think of it in the same terms as two of anything else smashing into each other. But galaxies aren't physical objects that work the same way as, say, two trucks colliding. Galaxies are collections of stars, and even in a superdense galaxy, there is a hell of a lot of empty space between each star. So if you get two of these things meeting, it's extremely unlikely that any of the stars will actually hit each other. Over the course of our galaxy's existence, it's actually already had a few dwarf galaxies pass through it. And like ghosts passing in the night, they just breezed right through.

But Andromeda is no dwarf galaxy. This is a serious business galaxy. Although there won't be much physical contact between the two galaxies, they are still going to fuck each other up quite a bit. Right now, all the stars in each galaxy are settled into orbit around the black hole that forms the galactic center. The stars have planets that orbit them, and the stars themselves orbit the black hole. Everything is pretty nicely balanced. But when a whole other galaxy gets mixed up in there, with tons of stars and its very own black hole, all that balance gets thrown off. The gravitational fields of each galaxy are going to distort one another, and the nice, orderly galaxies we're used to are going to get torn apart.

The extreme gravitational pull of each black hole will keep the galaxies from just continuing on past each other, and will eventually pull them together to form one big huge galaxy. The whole thing, from first smash to new equilibrium, will take millions of years.

The craziest thing about all this is that the Earth won't actually be disturbed by any of this madness. Current predictions say that our whole solar system will be unaffected. So for millions of years, future humans—if we still exist—will just get some spectacular views in the night sky.

ANDROMEDA ○ NASA/JPL-CALTECH

~~ANTIMATTER~~
(WAIT, WHAT?)

A deep dive with an ordinary dude
with guest writer Laura Pumphrey

Nature seems to love balance: day and night, hot and cold, decent human beings and huge dickheads—the list goes on. And it looks like nature's boner for balance extends right down to the most fundamental level. We're all familiar with matter (the "stuff" of the universe), but physicists have decided that to this ying there is also a yang: antimatter.

Here to explain some crazy shit about antimatter is Laura Pumphrey. Laura is not a physicist or any kind of legitimate scientist. She is a sculptor, a barista, and an aspiring farmer. She reads about the stuff that interests her, and now here she is to share her knowledge with you.

The case of the missing antimatter in our present universe is a bit of an eye-opener to say the least.

Auntie who? I know she matters, but what about uncles? Just because she lives in Calgary doesn't mean she's missing. We all matter—we ARE matter! Why are you so against me? What the Sam Heck is the meaning of all this?

Antimatter violently reacts with matter—the stuff of which our bodies, the Earth, the stars, the galaxies, and everything in the observable universe is made—in a fantastically explosive way. Antimatter and matter are true opposites, weird mirror images of one another. They have the exact same mass but opposite electric charges. Two opposing particles, say, for instance, a negatively charged electron and a positively charged antielectron (a positron), are attracted to each other. Like lovers in heat, they pounce on one another and WHAM! Contact happens! Two become gamma rays! A truly photasmic event. They destroy each other in a hot fit of destiny and create energy equal to the electron–positron rest mass. (Side note: if it really were just one electron and one positron kiss-off event, we wouldn't even notice, unless you had the superpower to sense elementary particles. Our bodies even produce about one positron every twelve seconds, and bananas squirt out about a positron every seventy-five minutes.) Ever had a PET scan? That means you've had tiny amounts of a radioactive tracer that emits antimatter injected into your bloodstream, or that you swallowed antimatter like a delicious juicy drink.

ANDREA HSIEH

X-RAY OF A BULLET CLUSTER ○ NASA/CXC/CFA/ M.MARKEVITCH ET AL.; OPTICAL: NASA/STSCI; MAGELLAN/UNIVERSITY OF ARIZONA/ D.CLOWE ET AL.

So far as we can see and detect things, when high-energy fluctuations happen, whenever matter is produced—in cosmic rays, high-energy flashers, blasters like the big bang—antimatter is also whipped out in equal parts, and they pounce on and annihilate one another. Producing different amounts of antimatter and matter from high-energy events is unheard of. Wait, what? Are you sure? The big bang made all of us. I feel like you're focusing on the auntie–uncle side of things again. Why was it supposed to create equal amounts of both matter and antimatter? That's too perfect. How do we know for sure? Well, a brilliant physicist named Paul Dirac, a prophet to a god he didn't believe to exist, was able to infer the existence of antimatter, before antimatter was even discovered to exist. His equations started out innocently enough. He was playing around with electrons traveling up to light speed, and according to the energy equation $E=mc^2$, he found that another solution kept popping up again and again. What? Two valid solutions to the energy equation? "What's this?" he must have asked himself in confusion. "A positively charged electron?" He apparently brushed it off at first, thought it might be a mistake, and continued on mathin'. Dirac finally stepped up to the plate and confidently presented his findings to all who would listen. Both scenarios in the equation's answer were acknowledged, and the energy solution was publicly allowed to be either positively or negatively charged. Two separate instances regarding one single particle with a charge we're familiar with. Dirac of course crunched the numbers for protons. He found there was a negatively charged particle as its counterpart, calling it an antiproton. Then the Nobel Prize waltzed into HIS life equation, taking up square footage not only in his imagination but right there in his favorite room that existed in real life.

Creating different amounts of antimatter and matter in energetic cosmic events has never been observed.

They're always just twinned up. We'd probably have to change some laws of conservation if that were to happen. Clearly, the fact that our universe exists in its glorious matter-full state should be testament to some law adjusting. Perhaps there was just a little more matter? A slight smidgen, a slant, a teensy imbalance? What if there was 99.9999999999999999% antimatter vs. 100% matter in the primordial energy? Quarks (being what protons and neutrons are made of) would have had to be present in larger quantities than their antiquark counterparts and enemies in order to match their destructive energies canceling each other out, then continuing on to interact with each other creating protons and neutrons. That infinitesimally small difference would then be enough for matter to shine in the billions of galaxies in our universe that we can see, not to mention what we can't see.

So why else should we be so concerned with searching for a substance that can blow us up?

Antimatter, in larger quantities, can be used in multiple terrifying ways by the likes of paranoid humans addicted to power and control. We must be wise and on the lookout for any slimy, shaky hands trying to snatch it greedily from the camp of bettering ourselves. The calculations say that one gram of antimatter reacting with one gram of ordinary matter is equivalent to around three times the amount of energy in the Hiroshima bomb—one gram! An excellent weapon for warfare indeed. Most people say not to worry—with current creation rates of antimatter at CERN it would take approximately one trillion years and take one zillion dollars to produce a gram, maybe more. Perfect, because I really hope we don't destroy the Earth and blow each other up. Imagine the potential it possesses as an energy source—maybe one day we'll be able to control the explosive fury into a form of energy that doesn't pipe out gigantic loads of carbon dioxide into our Earth's tender atmosphere.

Maybe we could even use it to travel through space reeaaaaaaally fast, close to light speed! In giant spaceark rocket ships. Containing biobots, DNA, seeds, and technological instruments able to test the surroundings of unknown environments. We could be an Earth seedling containing all earthly and planetary knowledge, armed with the intelligence to communicate peacefully and use good judgment in a way to better the journey for everyone, including fellow citizens of the cosmos. We could be observers on a universal exploration. With incredible giant electromagnetic vacuum-holding cells of antimatter, future Earthlings might figure out how to fuel their ships by harnessing and manipulating the annihilation of antimatter and matter for propulsion. Just like Star Trek's *Enterprise*.

Although what we see in space is predominantly matter, we can detect antimatter flooding out of galaxy centers like fluid jet-stream fountains, swirling around and unfurling in galactic clouds. It's produced in high-energy cosmic rays like those from supernovae—and even above strong bolts of lightning in thunderstorms! And, of course, inside our bodies. The physicists at CERN are amazing and found a way to make antihydrogen. No biggie, they just did some tinkering and built a decelerator because the accelerator wasn't good enough. We haven't seen any antihydrogen in space yet, nor any other element. The Alpha Magnetic Spectrometer on board the International Space Station is searching darn hard for antimatter of all kinds. No antimatter galaxies yet, which might actually be the most terrifying thing ever to observe. Theoretically, antielectrons can orbit antiprotons and antineutrons, leveling up to create anti-elements and quite possibly anti-planets inhabited by anti-feminists. Perhaps there're some creepy antimatter universes looming outside our own, ready to abominate our beautiful bubble with enough anti-everything to destroy every galaxy, sun, planet, and all of the dark stuff extremely violently. Imagine being able to detect regions of space that are attracted to each other at first and then begin to nightmarishly annihilate each other upon contact? How would you begin to explain the wild news to the people you love? *Hun, the antimatter is coming! You can run, but you sure as hell can't hide. What?! No, babe, we're all fucked. Yes, yeps, all dead—even the kids.*

It might bring some solace to be informed about a couple of experiments involving the decay of certain particles. Apparently the weak force is something that's a little hard of hearing and might have missed the boat in treating matter and antimatter equally. Some rules were broken in the United States in the 1960s that you might have heard about. Yeah, the weak force (when dealing with kaons in this particular experiment) is unsymmetrical under CP symmetry. Um, what did you say to me? The weak force, one of the four fundamental forces in the universe, probably doesn't treat antimatter and matter entirely the same. There's more experimenting to be done, but basically the weak force is a big bigot. The weak force recognizes that matter matters more than not matter, and it's like, "You're allowed in, you're allowed in. You! Get to the side so I can frisk you. HEY THERE! Wait a minute, buddy, where do you think you're scampering off to?!" It's just not that into antimatter; in the weak force's mind, as much as it really does matter in the ways that it

IS matter, it just doesn't. And the weak force is one of the universal bad-boy bouncers.

It's kind of a beautiful thing that the universe is leftover stuff that survived the great annihilation period of the big bang.

There must have been a lot of matter and antimatter coming out of god knows where. Kind of neat that it's asymmetrical. Asymmetry is much more interesting and beautiful than absolute perfect symmetry. Our Earth is a little squished in the middle. Are your hands the exact same? Everyone's got a crazy eye. A complete perfect balance isn't necessary; it would take away from the reality of the moment if we were constantly obsessing with perfection and symmetry. In art school they say asymmetry is more appealing than complete balance. Your brain be like "What?! That image is not exactly like the other one! This is aesthetic anarchy!" And you store away the image in your memory banks forever because you were so confused and ended up staring at the collage for so long. There's much more than just a few pieces of the universal puzzle missing, and if the force of this creation isn't perfectly symmetrical, that's A-OK with me—not as though I have a choice in the matter.

DARK MATTER
& DARK ENERGY

ALEX GVOJIC

DETAILED DARK MATTER MAP ● NASA/ESA/D. COE (JPL-ALTECH AND STSCL)

PEOPLE TODAY FRONT LIKE WE KNOW EVERYTHING. WE'VE SEQUENCED THE GENOME, HOOKED UP THE INTERNET, CURED DISEASES, FOUND BLACK HOLES, AND INVENTED BASKETBALL. WHAT MORE COULD THERE BE?

Well, it turns out there's a hell of a lot we don't know. We actually have no idea what makes up 96 percent of the universe. Which, you know, is a lot.

We know that we're missing this huge chunk of what makes up the universe because we see it having an effect on the stuff we do know about. It's like being able to see a shadow without being able to see what's casting it.

So far, we think that there are two different kinds of mysterious shadow-casters in the universe: dark matter and dark energy.

Dark matter is the term we use to tidy away a big problem: there seems to be a whole lot of mysterious stuff (AKA dark matter) that has a gravitational pull on the other stuff (regular, run-of-the-mill matter) in the universe. What's wild about dark matter is that it doesn't appear to be made of atoms. We figure this because we are pretty good at detecting atoms and we can't see this damn stuff. Yet it looks like there's six or seven times more of this dark matter than there is ordinary matter, and we have no fucking clue what it is.

The other spooky mystery of the universe right now is dark energy. Basically, scientists have known for a while now that the universe is expanding. They thought maybe it would expand, slow down, contract, and then do a "big crunch" back down into a singularity before doing another "big bang" and expanding again—going in a neat little cycle. Wouldn't that be nice, eh? Putting things into an eternal cycle answers questions about the origins of the universe, at least.

But scientists found that the universe's expansion is accelerating, and they have no good explanation for how it's doing that.

Imagine you're riding your bike on flat ground and suddenly start speeding up without pedaling. Either a ghost is pushing you or there are exceptions to the basic rules of physics. In the universe's case, most scientists prefer the ghost explanation. They don't want to deal with the idea that physics could be fundamentally wrong. So they say, "Well, there must be something pushing everything outward, and we just don't know what it is yet."

Maybe this is a cop-out, and we just don't actually have the formula for how the universe works. Or maybe the universe really is full of weird ghostly shit that we can't yet figure out. Either way, there's a hell of a lot of mystery left to contemplate.

CAL SINADINOVIC

"TWO POSSIBILITIES EXIST: EITHER WE ARE ALONE IN THE UNIVERSE OR WE ARE NOT. BOTH ARE EQUALLY TERRIFYING."

—ARTHUR C. CLARKE

Aliens are out there. You already know the universe is big as fuck. Statistically, it's highly improbable that the only life is terrestrial. That's not really disputed in the scientific community anymore. There are just too many worlds for this to be the only one to have sprouted life.

There are scientists whose whole job is to figure out just how likely alien life is. This badass career path is called exobiology, and it's given us some impressive numbers to contemplate.

TREVOR BAIRD

We know from our time on this little pale blue dot that life is impressively diverse, so it's tough to even begin to imagine what kind of insane life-forms might exist in the dizzying array of conditions this universe has to offer. We don't know what it would look like, we don't know what it would be made of, and we don't even know if we could see it if it were right in front of us.

Now, finding this stuff is a pretty daunting task. But we've got leads!

We can say with certainty there is life on Earth. So we know that it is possible for life to develop in the particular set of conditions that Earth has. If we're going to find life in the universe, our best bet is to find places just like home. So that's where the exobiologists start.

They look for stars similar to our Sun, with planets orbiting at distances similar to Earth's. Planets under these conditions could have liquid water, which is the one thing on which all Earth-bound life depends. And even with these stringent conditions, it turns out there are tens of billions of those planets in the Milky Way alone.

If we expand this to the whole universe, we get about 6 quintillion potentially habitable worlds. "What the fuck does that mean?" the discerning reader asks. Well, it's a shitload. With a pretty damn conservative guess that only one in a billion of these worlds will contain life, we still wind up with over 6 billion alien-infested planets. So that's it. It's basically certain that life exists on other planets.

Of course, this doesn't mean that every single one of them will have full-blown biodiversity to the extent that we do. Earth spent most of its inhabited history covered in microbes, and microbes alone. Chances are most alien worlds are just covered in alien bacteria. (Uh . . . gross.)

The average person doesn't give too much of a shit about microbial life on this planet, let alone any other one. We're better at wrapping our heads around little green men or telepathic super geniuses: aliens who might someday come here to blow us up or teach us the secrets of interstellar space travel.

The existence of creatures like these is pretty damn unlikely. But not impossible!

ALIEN LIFE MUST EXIST IN THE UNIVERSE.

Many space scientists are sure that intelligent, technologically advanced alien life must exist in the universe. They base this on a similar numbers game to the one we played earlier. In all likelihood, creepy aliens like that are too far away from us for communication to take place, either intentionally or by accident. But they're probably out there.

When we think of these kinds of aliens, we tend to think of the hyper-intelligent alien scientists and engineers that build alien spaceships and explore alien planets. Though they might be out there, forget for a moment about the aliens that science fiction chooses to focus on.

Think, instead, about the alien deadbeats. The losers who coast along in their advanced alien society, living in their alien parents' alien basement, sleeping until alien 2:00 p.m., and spending the rest of the alien day watching the alien *Bachelorette* and getting fucked up in whatever way those aliens do. Take comfort in the notion that somewhere out there, in the distant reaches of the universe, there's an alien as lazy as you are.

By some stroke of cosmic good luck (God? Are you real?) we live in a universe where electromagnetic radiation is a thing.

....

Electromagnetic radiation is basically energy that moves through space in a wave form. Light is electromagnetic radiation, but there are a lot of other kinds that we can't see. Microwaves, X-rays, radio waves—these are all kinds of electromagnetic radiation, and we've learned to mess around with them for various purposes. Cook your food by putting it in a box? Sure. Look through flesh to see broken bones? Yeah, whatever; we live in the future, we can do that. Communicate instantaneously over long distances? It's fine, we got this. We have figured out how to manipulate energy to do all these things, and that is pretty fucking cool.

We could write a whole book called *Electromagnetic Radiation and Its Applications Are Cool as Fuck*. And believe us, you'd want to read that book. But here we are, writing a space book, so we'll just talk about a few cool things.

Cool thing number one: communicating at the speed of light.

Humans have made massive leaps in communication over the last few centuries. In less than two hundred years, we went from having to physically carry a message from one place to another on paper to being able to video chat. (Even we, the enlightened authors, forget this sometimes and bitch when we have a sketchy connection on Skype.)

You can transmit your voice, send your text, or fly your RC helicopter all because information can be conveyed by radio wave.

It gets easier to understand if you think about communicating with the part of the EM spectrum that we can see: light. Imagine communicating with someone by flashing lights back and forth using a code that has to do with how bright the light is or what color it is. Well, you can do the same thing with other kinds of radiation by adjusting their amplitude (that's what brightness is!) or frequency (that's what color is!). There are other properties of a radio wave that you can fuck with too, so you can pack a lot of information into a radio signal.

....

You can transmit your voice, send your text, or fly your RC helicopter all because information can be conveyed by radio wave.

....

You just need to establish the code that you're working with and program the sending and receiving gadgets to understand the same code.

This is how cell phones, radios, TVs, and a shitload of other things work. And it's also a massively useful technology in space. Everything we have in space would be totally useless without radio communication, because it's impossible to make physical contact with them. We shoot invisible waves of energy at our spacecraft, and they do what we tell them. They then send us back invisible waves of energy, we catch them in big radio dishes around the world, and then we decode them. Every stunning Hubble image you've seen and every picture you have in your head of Jupiter or Saturn all came to us in the form of waves of energy whose properties we decoded.

Electromagnetic radiation has another extremely sweet application in space: spectroscopy. When you shoot radiation at something, some of it bounces back at you, and some of it gets absorbed by that thing. What sticks and what bounces back depends on the properties of that thing. We're all familiar with this in the case of light. For example, blue things look blue because they absorb most of light's radiation but let radiation of a particular frequency bounce off. When that bounced light hits our eyes, it's decoded as blue.

Well, this happens in parts of the EM spectrum other than visible light as well. You can get a lot of information about something from shooting radiation at it and seeing what bounces back.

Scientists have messed around with this technique enough to know how radiation interacts with just about every substance that exists. We basically have an encyclopedia of what they call "spectral signatures" and when we want to figure out what something is made of, we can look at how radiation bounces off it and use that encyclopedia to find what it is.

TOP: RIA MCILWRAITH ○ **BOTTOM:** ANDREA HSIEH

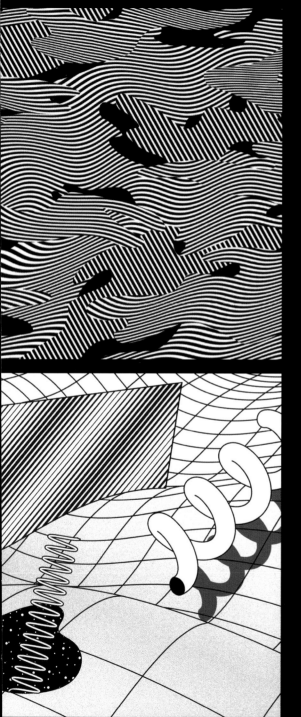

This is particularly useful when deciphering what is going on out in space. Scientists can look at the radiation coming from another star and figure out how big it is, what gases it's made of, how hot it is, and more. We can look at something like the radiation bouncing off Titan's seas and figure out their chemical composition. And we could conceivably look at the radiation bouncing off the atmosphere of some distant planet and figure out that there are gases present that must have been produced by alien farts.

It is truly cool as fuck that just by understanding how energy moves in waves, we can do all these kinds of magic.

· ·

Lots of people hear that word "radiation" and immediately think of nuclear fallout and mutants. But that "bad" radiation is only one specific part of the spectrum of radiation. Light is radiation! It's not all bad.

· ·

~ **CHECK THIS SHIT OUT** ~

The fastest pulsar, a kind of neutron star, spins 716 times per second.

BILL NYE

knows exactly how fucking cool space is!

BY SOME WIZARDLY STROKE OF MAGIC, BILL NYE AGREED TO SIT DOWN WITH US AND ANSWER ALL THE QUESTIONS WE COULD THROW AT HIM ABOUT SPACE, SCIENCE, AND THE FUTURE. WE EVEN GOT HIM TO SWEAR.

Our first question is something that is probably Google-able, but we want a better answer for it.

How did the universe start? I mean we know there was the big bang . . .
Why do we know there was a big bang? What makes people say that? The reason people say there's a big bang is because we observe all these stars moving apart. And this is just a hugely compelling idea because its consequence is that everything that we can see, including us, was in a volume that is, I think for most of us, literally unimaginably small. And in 10 to the minus 43rd seconds, all of that started to expand. And you know why? Nobody knows why. So really, it's a crazy thing.

I think the bigger question for me and I think a lot of people is . . .
Bigger than the big bang, a question bigger than the big bang . . . ?

Bigger than the big bang! There's this idea that the big bang started with this tiny, tiny point that everything expanded from—but where did that point come from?
Nobody knows, man.

Nobody knows.
Is there a god that directed all this? Why did she do that? I don't know, man.

Dare you speculate? Do you have theories that you like? Is there mathematical evidence of some sort?
I have no idea what came before the big bang, but if you haven't wondered about it, I don't know what you do with your time. I hope you've wondered about it. When I was a kid, it was speculated that the universe might expand and expand and expand and expand and then collapse again.

Yeah. I liked that idea. It was reassuring, the idea that we're just in this infinite thing going back and forth.
But it doesn't work like that. It doesn't look like that's how it goes down. And then people speculated that the universe was uniquely, mathematically, asymptotically flat. Where it would expand, expand, expand, expand, but never stop expanding. And you determine that by observing distant stars. But these guys set up this worldwide network to observe supernovae, and they determined, conclusively, that the universe is not expanding—it's accelerating. Nobody knows why that is . . .

So it's gonna move toward, what is it, a big rip?
That's one theory . . . you see, everybody's fascinated with the astrophysics; everybody wants to know these answers. Meanwhile, we have climate change, 7.2 billion people, is this guy you're dating going to marry you?

BILL NYE INTERVIEW ☾ 65

BRETT RANDALL

Those are important questions that are sort of more everyday and more important to your everyday experience than whether or not the universe's expansion is repeatable or is there gonna be a rip . . . is there gonna be more than one universe, a multiverse, or five, or thousands . . . ?

It's those questions that when you're not worrying about your boyfriend and you put your smartphone down for a minute, yeah, everybody wants to know about those.

I find it makes my general day-to-day experience a lot nicer, thinking about the big bang stuff.
Oh man, absolutely.

Next question: Are we fucked? Seriously. Have we ruined the environment, is it too late, is there any turning around, is there any mitigating, are you hopeful?
You have to be optimistic. If you're not optimistic, you ain't gonna do no nothin'. However, we have put so much carbon dioxide in the air so quickly that a lot of us are fucked, if I understand your question.

So what we've got to do, as I always describe it, is everything all at once. In other words, you want to do as many things as you can as quickly as possible to deal with climate change. There are people that are quite sanguine about our situation in regards to climate, and I don't think they should be.

This is to say that at no time in the last 680,000 years, the so-called seven interglacial periods between glaciation of the Earth, has the carbon dioxide level gone above 280 parts per million. Until 2015, when it went over 400.

The thing about carbon dioxide: not only does it keep the world warm, it also doesn't break down. Carbon dioxide that is already in the air is gonna be there for centuries, so there are schemes being proposed to take the carbon out of the air, but it's really difficult. You wanna take out about an eighth of what's there. That's a lot of carbon dioxide, quintillions of tons.

However, the longest journey begins with a single step, and I am open-minded about nuclear power if somebody is not crazy. In other words, if the nuclear industry could be trusted to not build plants on Fukushima-style earthquake faults, to not have Chernobyl explosions, and to not have Three Mile Island "You guys go this way; we'll all go this way because the valve has got a scratch in it." The nuclear industry just hasn't been that sharp, but if they could tighten it up, I'm open-minded.

Do you think it's possible to cut down how much carbon dioxide we produce or actually reverse some of the damage that's been done?
Reversing damage, absolutely. You can reverse damage like crazy. And the first thing you would do is motivate societies and governments to not cut down trees without afforestation, as it's called: planting new trees. It has a huge effect. It doesn't take care of everything, but it's a huge effect.

The big thing I would like people to do is to think of the Earth as a house. When you are a renter, you call the landlord when stuff goes wrong: plumbing, electricity, etc. Some people even call the landlord when the lightbulb goes out; that's pretty hardcore. But if you own the house, you can't do that: you're the landlord, you have to fix the furnace, you have to repaint it. What kind of paint do you use, what color's it gonna be? *Should I fix the plumbing now, can I afford to fix the plumbing? Do I want air conditioning, how cold should I make it and how much can I afford to pay, how much electricity can I afford to buy?* All these things become your everyday continual decisions.

So my big push is to get humankind to think of the Earth as your house. It's so hip to say, "It's our home, and I want to make a house a home." And that's right, I'm talking about accounting, talking about keeping track of the numbers. I think if we all did that, we'd think a bit differently. There is also this term "geoengineering": engineering the whole Earth. Some people say we should pump sulfur dioxide into the atmosphere, just like volcanoes, to reflect sunlight into space because that's what volcanoes do. But we'd get acid rain, and that seems like kind of a bad thing. "OK," they say, "we're gonna make mirrors and put them out between the Earth and the Sun." To that I say, are you high? I mean, to launch any rocket is hundreds of millions of dollars, or euros, or whatever, and they're just not that reliable. And where are you gonna put it, and who's gonna be in charge of that? And so on. Then people want to have artificial trees to take carbon dioxide out of the air, but who's gonna manufacture that, who's gonna take charge of that?

But there are a couple of ideas that I find really appealing. One is to preserve the ice. Do what you can to preserve the ice in the Arctic or Antarctic or whatever, to reflect

NEECHO

more sunlight into space. If you can invent pale pavement, instead of black asphalt, some kind of white, you'll be rich! And not just fill it with concrete—it's gotta be as cheap as asphalt. It's not an easy thing.

And then the big thing that would change every everything everywhere is a better battery. Or it's called a system of batteries or a family of batteries: four or five different kinds of batteries, each has its own perfect application. If we could store electricity at night when the Sun's not shining and during the middle of the day when the wind's not blowing, you could really pull that off.

So everything all at once.

And, as we say all the time to conservatives, or climate-change deniers, "You don't like regulations now? Just wait until climate change really hits, when the drought in California gets worse and worse." There's people who live in the Miami area who can't get liability insurance for their cars because when there's a storm their cars get salt water up to the doorsills and it just corrodes the cars, so they don't insure it. And that's climate change. As the ocean gets a little bit warmer, it expands, water expands, and the ocean's gonna come up closer and closer to the shore. And yes in the developed world we'll build sea walls at ports, but in the developing world people aren't gonna have that luxury: they're gonna be displaced, there's gonna be refugees, there's gonna be trouble.

Regulations are gonna happen. My parents both were in World War II. And my dad was a POW, and my mom was in the US Navy, as a photographer. But you couldn't buy tires at a certain time of the year, you could only buy a certain amount of gasoline every week, you couldn't get sugar and salt—only at certain times when it was rationed. And people did it, they did all that. And they changed the course of history by defeating some crazy guys.

So humankind is capable of great things. My parents' generation was called the greatest generation because they did this. I want the next great generation—I want you people—to step up.

OK, a couple of space questions—it's a book about space, after all. One of them is pretty straightforward: do you think you'll get to go to space in your lifetime?
I think I'll have the option, but I don't think I'll take it. By that I mean if the price point stays at $200,000 for three minutes in space, I probably won't do it. But what if I'm diagnosed with cancer? I might take some of my treasure and blow it on a ride at $50,000 up and down for three minutes and take some pictures and get that global perspective, which apparently changes astronauts' lives. After they see the Earth from space, their perspective is changed.

Something that a lot of people are interested in is the colonization of space. There's this idea that humans are inevitably gonna move toward colonizing other planets; NASA's talking about learning to settle on the Moon or Mars, etc. What's the point?
Well, you got the wrong guy for that.

This is what I tell people: You really think you want to live on Mars? You're crazy. I really want you to reconsider that. Go live in Antarctica. If you want to know what it's like on Mars, start with Antarctica, and don't go to the shore where the birds are flying around and penguins are getting eaten by the orcas or whatever happens; huge whales are eating krill—no. Go to the dry valleys. It hasn't snowed or had any precipitation for over a century. And don't breathe: take all the scuba tanks you need for a couple of years and see what you think. You will not like it; it's a drag.

But there's generally been this idea, this push where you've got to move toward human presence on other planets as though it is our duty to expand into the universe. Do you think that's legitimate?

In the United States, we have an expression that dates back to the 1800s, "our manifest destiny." And I think it's a European point of view. These people came from Europe, where conditions were not nearly as good as they could tell that they should be. So they just kept going west, to North America, and then from the East Coast just kept going west.

And people just had this expectation that you just keep moving and keep shooting things and eating them, and growing things and eating them . . . do you know the state motto of California?

No.

"Eureka, I found it." These are European guys who come over the hill, and oranges are like weeds, and in the Sacramento River there are salmon as big as your torso just swimming there into your lap. And so people just have this expectation of Mars that is completely unreasonable.

The Earth is it, the Earth is our house. You've got to do the accounting, you've got to grow the food, we've got to take care of each other here in our house.

Yes we have to explore Mars, we have to explore Europa, and the reason we do that is to look for signs of life. I love rocks; some of my best friends are geologists. But I want us to focus on looking for life. Rocks help you find water, but let's take it that next step. And we don't have to spend that much money to do it. If we were to discover life elsewhere, it would change this world.

I've heard you say that a number of times. I wonder, do you have specific ideas of how the world would change? What would be different?
I think when you find out that there's life on another world, I hope you ask the next question: Is it like us? Does it have DNA? Are we the descendants of Martian microbes—which is not unreasonable? It's extraordinary, but it's not crazy. It would change the way you think about what it means to be a living thing in the cosmos, and I hope it would change the way you thought about the Earth.

The Martian microbes, if any have managed to survive in some icy gully near the equator, they're not doing as well as Earthlings, with all our fish and oceans and phytoplankton and redwood trees and what have you. So you want everybody in the world to appreciate that. Now if it turns out you can prove conclusively that there never was life on Mars, that would be troubling in a different way.

It would just be one definitively lifeless planet, though.
That's it. Even Mars, which has such great promise as a once-wet world, at a certain distance from the Sun, which used to have an atmosphere—if even on that planet there is absolutely no chance of something alive, that would also be a chin stroker.

Do you think that it's likely that anytime soon we'll detect a signal from an extraterrestrial intelligence?
It's as likely as not. And I tell everybody, there's one way to make sure you do not hear a signal and it's to not listen. If you play hockey, you miss a hundred percent of the shots you don't take. So you have to listen. This is not to say that society drops everything we're doing and builds all our artificial intelligence machines to listen for a signal. Just listen. And if we got a signal it would change the world.

And now with these discoveries of Earth-like planets or planets in habitable zones of other stars, that's where you go listen, that's where you point, hoping you experience that there's something out there. Man, that would be something else, wouldn't it?

Yeah, I agree. It would be crazy, and it would change everything in a very different way. Especially because those people we detect a signal from, or those beings, would likely have been sending it a very long time ago, and who knows where they're at now.

OK, moving on. Mathematicians often talk about this exquisite feeling that comes with reaching a greater understanding and when they find a solution to a problem. Have you felt this often, and is it accessible to the rest of us?
Yes, absolutely. I'll give you an example. You design a Halloween costume, you have an idea for a Halloween costume, and you've spent a lot of energy on it—bet you've done this at least once. And you get the reaction you wanted from all the people at the party or who come to the door [or whose door you go to], and you feel great about that. You designed something, you made a prediction, and it worked.

So when you discover mathematical phenomena—you predict how long the candle will burn before it goes out, when you determine how far you can drive the car before it goes empty and it works out—these things stick with you, they're wonderful.

There's amazing subtleness in the universe described by mathematics that gives these people, they claim, an extra insight. I definitely go along with that. As I like to say, as an engineer who went into this nature-of-control stuff, the world is black and white until you get to second order differential equations. The whole freaking world is described by second order differential equations. Now do I have an insight into catching flying discs that other people don't have? Probably not. I mean, I've seen an analysis of flying discs that's pretty freaking complicated as differential equations. So it's some of each I guess.

When you prepare a recipe that comes out right, that's a great feeling.

So it's the same feeling; they're just calling it elegant?
Well, it's elegant and it's—well, these people can put a spacecraft on Mars. It's amazing!

Maybe it's a spectrum?
That's what I'm saying, it's a spectrum. It's so difficult. You experiment with a recipe, and it comes out better. It's really compelling.

So these people come up with mathematical equations of phenomena they observe in nature, and it works out. It really is something when you realize that no matter how hard you work, the number pi never repeats itself . . . and then that number shows up everywhere. It shows up in circles, the paths of planets, the diameter of trees, it shows up in statistics of assessing numbers of things or quantities of things having really nothing to [do with drawing] a circle. Everything that goes back and forth in this thing called sine wave: back there, there's a circle and the number pi somewhere. It is elegant.

So a lot of people that I know, who maybe are not super science educated but very firmly believe that aliens have come to Earth, say, "How did we get the pyramids, man?" and that kind of thing.
They haven't been exposed to enough geometry. The pyramids are square, and you can make a square with extraordinary precision using triangles. The famous three numbers, anyone? Three, four, five. You have a triangle that's three chain lengths on one side, four chain lengths on the second side, and five chain lengths on the longer side. One of those angles, between the three and the four, will be a right angle. Every freaking perfect time. So the ancient Egyptians worked that out, and they put the time and effort and energy into it to make it as square and nice as they could.

And they had a lot of slaves.
Apparently [archaeologists have] evidence that everyone was just into it. In the same way that some guys wanted to work and build Grand Coulis Dam. There's evidence that they just wanted the pyramids and society supported it. They fed them, gave them tents, living quarters, and there were marketplaces developed for the construction workers. And this is [from] the guy that worked on the *Science Guy* show, Ian Saunders, who now works at Microsoft. Anyway, he'd found all these historical records, and it was a seasonal thing. When it was time to harvest the crops, nobody was building the pyramids.

That's really cool. That's much better than slaves.
Yeah, it's very cool, but it makes for great movies when they're going to crush a guy for the sake of building a pyramid. The pyramids all had secret entrances. Tomb robbers all found their way into every one of them. They took it pretty seriously, but it shows you that maybe it wasn't all that mean-spirited, that's all.

So overall, when people have a tendency to say, "How could humans have possibly done this? This makes no sense; aliens must have been here," maybe it's similar to, "Oh my god, how did this tree come to be? God must have made it!" It's just not understanding it and wanting to attribute it to something external.
It's also selling humans short. I just think humans don't suck.

This is a big, secular human message. Humans aren't inherently bad; you're not born a sinner in the secular human point of view. You're not born owing the world something. However, if you want to be popular among your tribesmen, you better not be just ripping them off all day.

This leads to this question of altruism: that if you're not altruistic, if you didn't help people out, then people don't want you around. If you were a raving murderer, they'd kill you because they can't afford to have you around. This is the secular humanist point of view.

And the big thing about evolution which is so hard for everybody is that when humans do something, like build a pyramid, there's top down. Somebody was in charge. There was the head rock guy; there was a geometer who did the three, four, five triangles. There was the guy who found the best grain to bury with the pharaoh or whatever.

But in nature, that's not how it works; in evolution, it's bottom up. As the saying goes, the good designs eat the bad designs, and here we are. You and I are the product of four and a half billion years of experiments, and it's bottom up; the modern world is organic. It's very different. Someone was in charge of the pyramid. When it comes to deciding whether you have five fingers or six, that just happened. And the fives did pretty well, so here we are.

Any parting words of wisdom for us youngsters who want to change the world and be good people?
You *say* you're youngsters; you're adults now.

We're young adults.
There are three things.

First of all, everything each of us does affects everyone in the whole world, which seems incredible at first until you think of the atmosphere. There's only one atmosphere, so everybody's sharing that. We all share the air.

Then as my father said, everybody is responsible for his or her own actions. You've got to go into it. Everything you do, you have to take responsibility for it, and when you mess up, you've got to say, "I messed up."

And the big goal is that we want to leave the Earth better than we found it; that's what you want to do. And this is definitely deep within us. You want to pass on, you want your legacy to be better than your predecessor's. You want to leave a better hand than the one you were dealt.

And what's the best thing to do to make the world better? I don't know, man. Figure it out! You're the guys in there texting with both thumbs.

Do you have a favorite song about space?
I like "The Universe," Monty Python's "The Universe." Big fan of that. But I mean, I like the Planets. I like the *Star Trek* theme. So I guess there isn't one that I listen to over and over. Oh, you know what I really like is "Moon Girl." See, there was a time

when everyone knew the phase of the Moon because there were no streetlights—or no lights—and your whole life you're always taking that into account. And now we take a whole bunch of other stuff into account.

Can you give us a sound bite? I think we may have told you that the name of the book is *Space Is Cool as Fuck,* **so we're trying to appeal to that young generation of people. So I would love, Bill, to get you saying, "Whoa, space is cool as fuck."** Whoa, space is cool as fuck! ⌇

ALEXANDRA ENGLISH

HOW ROCKETS WORK & WHY THEY SUCK

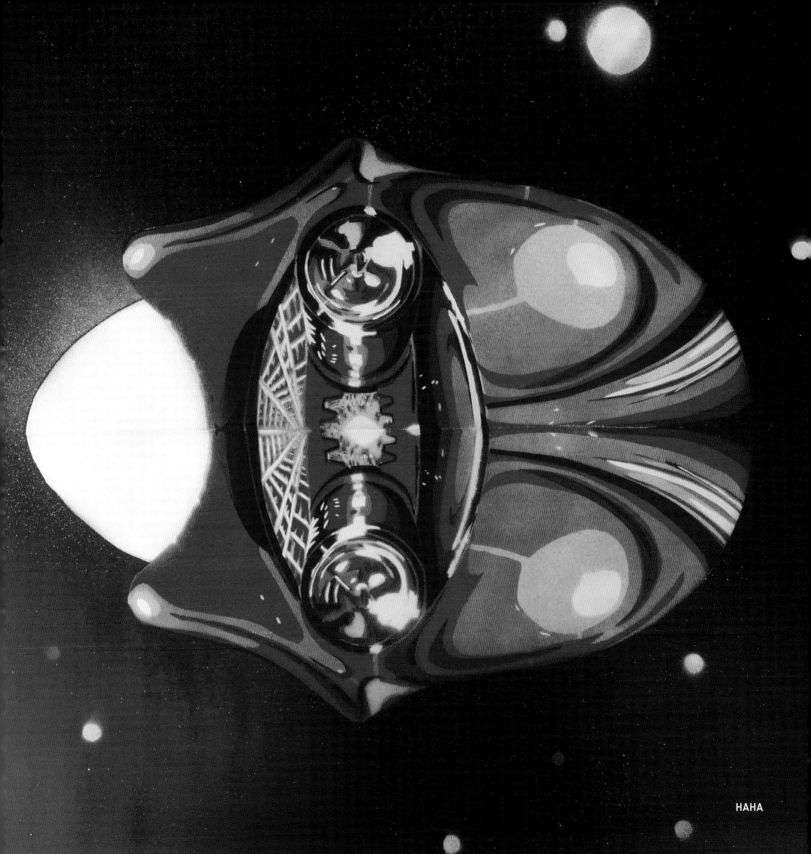

BLAST

AT THIS POINT IN TIME, WE EARTHLINGS HAVE ONLY ONE WAY OF GETTING STUFF OFF THE PLANET. WE STICK IT ON TOP OF A BIG CYLINDER OF FUEL, SET THE FUEL ON FIRE, AND BLAST IT AWAY FROM THE GROUND UNTIL IT'S FAR ENOUGH FROM EARTH TO BE UNBOUND BY ITS GRAVITY.

IT'S UNDENIABLY IMPRESSIVE THAT WE CAN EVEN DO THIS, ESPECIALLY WITH THE LEVEL OF PRECISION WE CAN ATTAIN. BUT NONETHELESS, ROCKETS KIND OF SUCK.

OFF!

A rather shitty aspect of this system is that the vast majority of the weight of the rocket comes from the fuel itself. We need a lot of fuel to get that weight off the ground, so most of the fuel we put into a rocket goes toward lifting the rest of the fuel.

What this all boils down to is that it takes a huge amount of fuel to get anything off the ground. And as I'm sure you can guess, rocket fuel is expensive shit. These days, it costs between $5,000 and $15,000 per kilogram (or about $2,300 to $6,800 per pound) to get stuff into orbit. The cost can depend on stuff like if you're sharing the rocket or where the rocket comes from (Ukraine has some cheap ones!). But in general, the cost stays in that astronomical range—no pun intended.

Think about that. A kilogram is about how much a typical pineapple weighs. To send a pineapple to space would cost around ten grand. And, of course, the price gets crazy when you're looking to send a big, metal satellite up—and even crazier when you want to also launch people and the shit they need to stay alive. A single Space Shuttle launch used to cost $500 million. What the fuck?

This is part of the reason we don't have space hotels, moon bases, global satellite internet, and other wonders that scientists at the start of the space age predicted we'd have by now. We haven't overcome this one massive hurdle of figuring out how to get stuff into space other than by setting a tube of fuel on fire.

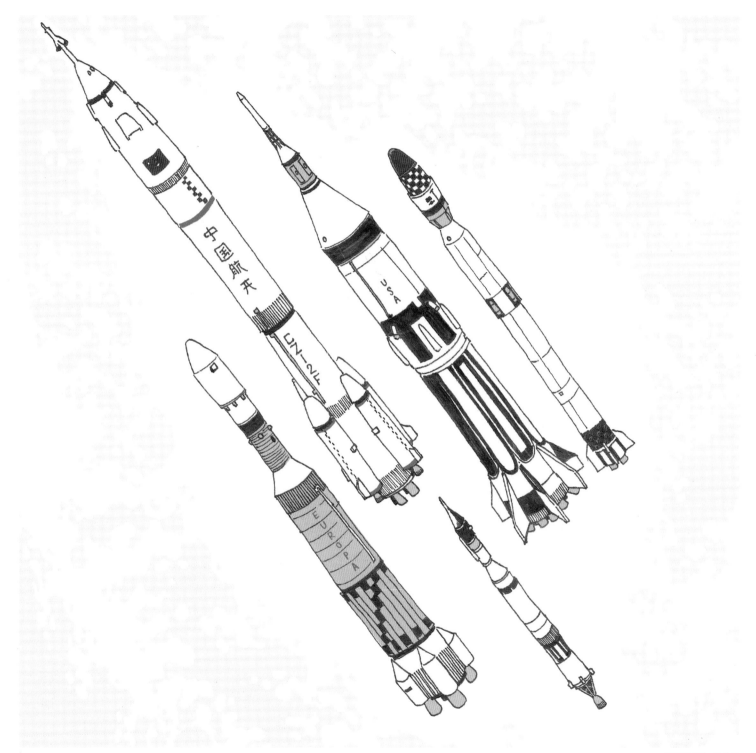

GORDON AULD

SPACE STATIONS

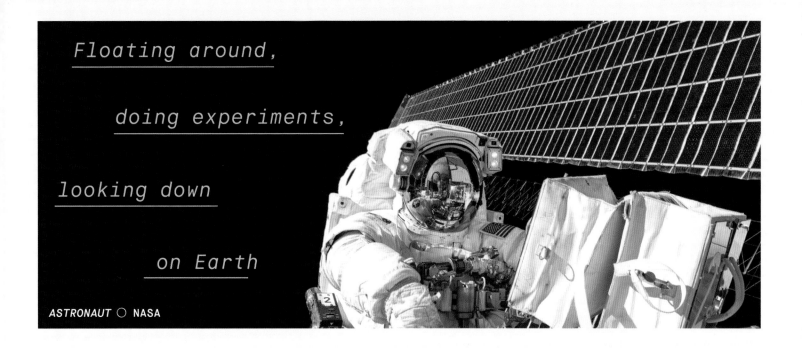

Floating around,

doing experiments,

looking down

on Earth

ASTRONAUT ○ NASA

YOU ARE PROBABLY AWARE, IN SOME CORNER OF YOUR BRAIN, THAT THERE IS SOMETHING CALLED THE INTERNATIONAL SPACE STATION.

YOU MAY BE VAGUELY CONSCIOUS OF THE FACT THAT THERE ARE ASTRONAUTS LIVING IN IT—IN SPACE—RIGHT NOW. BUT CHANCES ARE YOU HAVEN'T SAT DOWN AND CONTEMPLATED THE FUCKING MAJESTY OF THIS FACT.

LET'S MEDITATE ON THIS FOR A MINUTE.

Right now, as you read this, there are people in space (unless something tragic has occurred since the writing of this book) floating around, doing experiments, looking down on Earth with tears of wonder in their eyes.

The ISS has been up there and occupied constantly since 2000. And before that, there were smaller space stations run by individual nations. The first one went up in 1971, which is fucking nuts. Ever since people were strutting around in bellbottoms and watching *Soul Train*, we've known how to build structures that orbit the planet and can keep teams of people alive in space for extended periods of time.

The astronauts who live on space stations have kind of a weird time, no doubt. Some are up there for as long as a year, floating around in this huge, complicated box, shitting into bags (with no gravity— think about that), and doing experiments. It's been going on so long that it almost seems mundane.

There's never been a fatal disaster on a space station, though there have been some Hollywood-worthy close calls. And because it's not deadly or exciting or new anymore, nobody really pays attention to what's going on up there. But astronauts on space stations experience some freaky shit.

International Space

The ISS orbits Earth once every hour and a half, which means astronauts see fifteen sunrises and sunsets every day. If they're outside the station when the Sun peeks out from behind Earth, the temperature can suddenly swing from about -157°C to about +121°C (-250°F to 250°F).

Like anybody else, astronauts produce a bunch of waste. They pee, poo, barf, throw away their crusts, whatever. The big difference between you and them is that all their trash gets loaded into an empty spaceship and sent back to Earth to burn up into the atmosphere. So next time you see a shooting star, just remember that it might actually be a huge capsule of shit and garbage blazing down from space.

Astronauts don't have sex in space for a bunch of legit reasons, including the fact that astronauts on the space station are stuck together for months at a time and have to get along. Add sexy time to that and you might get problems. NASA and the other space agencies running the ISS have no official policy on space masturbation, choosing instead to just pretend their astronauts are superhumans.

If you cry in space, the tears don't fall down your face because there's no gravity. So they just create pools that surround your eyes. Extra sad! ▪

Station

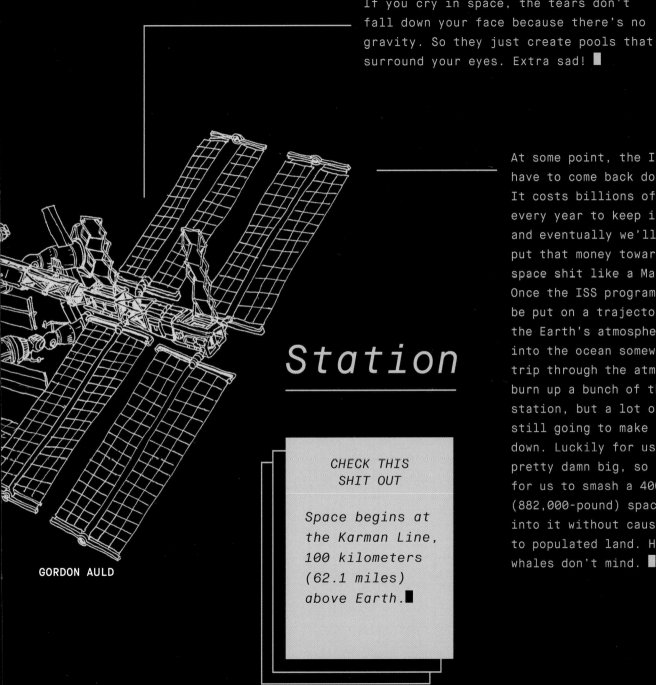

GORDON AULD

At some point, the ISS is going to have to come back down to Earth. It costs billions of dollars every year to keep it running, and eventually we'll want to put that money toward other cool space shit like a Mars colony. Once the ISS program ends, it'll be put on a trajectory to reenter the Earth's atmosphere and crash into the ocean somewhere. The trip through the atmosphere will burn up a bunch of the actual station, but a lot of debris is still going to make it all the way down. Luckily for us, the ocean is pretty damn big, so it's possible for us to smash a 400,000-kilogram (882,000-pound) space station into it without causing any damage to populated land. Hopefully the whales don't mind. ▪

> **CHECK THIS SHIT OUT**
>
> Space begins at the Karman Line, 100 kilometers (62.1 miles) above Earth.▪

JUST HOW BIG STUFF REALLY IS

YOUR BELOVED AUTHOR LIVES IN CANADA, WHICH IS A PRETTY DAMN BIG COUNTRY. IT TAKES A DAY AND A HALF JUST TO DRIVE ACROSS ONTARIO, FOR CRYING OUT LOUD.

But your author, dear reader, has a cosmic perspective, so she knows that Ontario is chump change compared with space bigness.

Jupiter is the biggest planet in our solar system by a long shot. It's so enormous that inside it you could fit every other planet in the solar system, plus all their moons, plus all the comets, plus all the asteroids. That's fucking huge, right?

But that's nothing in space. The Sun is almost a thousand times bigger than that, and ours is actually a sort of puny star. The biggest star we know of, UY Scuti, is five billion times bigger in volume than our Sun. FIVE BILLION TIMES BIGGER.

IT'S IMPOSSIBLE FOR HUMAN BEINGS TO EVEN FATHOM THAT SHIT.

But even that is nothing compared with the size of all of the nothing in space. Even in galaxies, which are relatively dense collections of stars held together by gravity, the average distance between stars is about five light-years, or forty-eight trillion kilometers. The average distance between galaxies: thirteen billion light-years. I'm not even going to bother telling you what that is in kilometers because it's a number too big to mean anything to you.

It's also pretty fun to try to wrap your head around the fact that even goliath stars like UY Scuti are entirely composed of unbelievably tiny particles, just like we are. Compared to those quarks, even us tiny, insignificant Earthlings are fucking enormous.

CHECK THIS SHIT OUT

SPACE IS SILENT. SOUND WAVES NEED SOMETHING TO TRAVEL THROUGH, AND SINCE SPACE HAS NO ATMOSPHERE, THERE IS NOWHERE FOR THE SOUND TO GO. SO A SPACESHIP FIRING UP ENGINES OR SHOOTING LASERS IN SPACE WOULDN'T MAKE A SOUND!

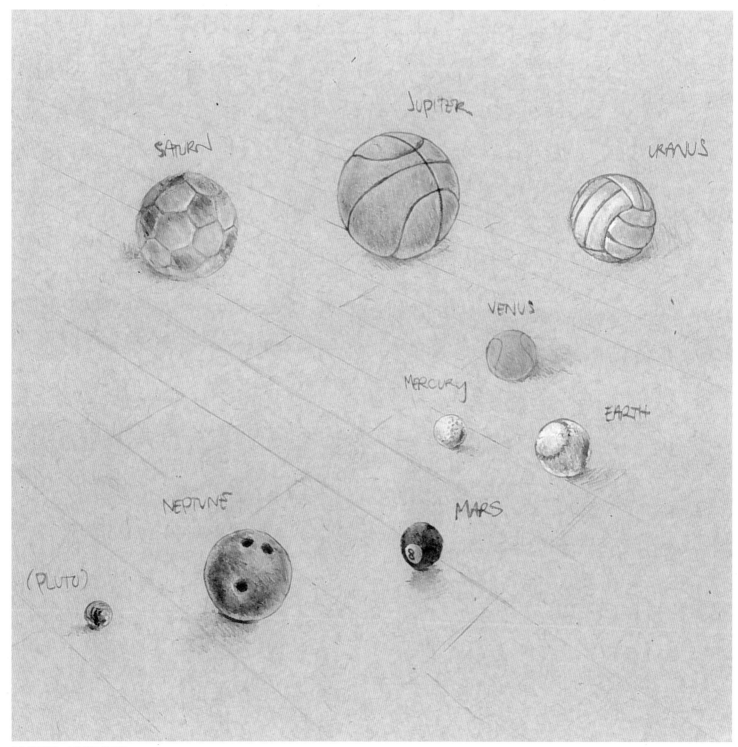

RICHARD AMEZQUITA

ALICE DOLLING

☾

THE MOON

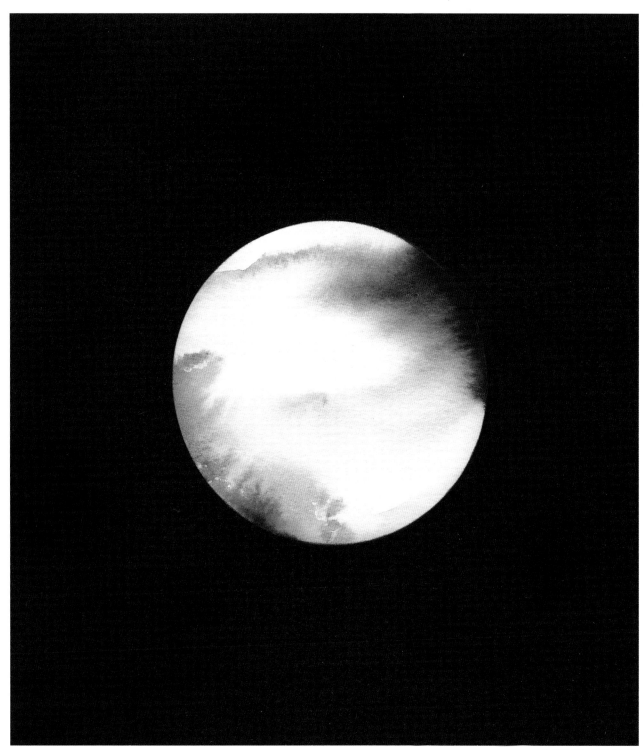

LAUREN BRADLEY

There was once this planet-like chunk of mass called Theia, so scientists think. It was about the size of Mars, and around 4.5 billion years ago, it was flying very quickly toward Earth. With no rag-tag team of drillers-turned-astronauts to divert it, it slammed the fuck out of our planet, caused a huge explosion, and sprayed a bunch of debris out into space.

The debris orbited Earth and, over the course of a few million years, slowly clumped together into a nice little ball. That nice little ball became Earth's only natural satellite, also known as the Moon.

There's plenty of evidence for the appropriately named "giant-impact hypothesis" that we can plainly see on the Moon's surface. Roughly 60 percent of the Moon is covered by dried lava that came from a bunch of rocks colliding at very high speeds.

Those rock collisions aren't only from the Moon's creation, though. They also result from the countless asteroids, comets, and other destructive shit that have pelted the Moon's surface over the past few billion years. The bumpy mosaic of violent, cosmological collisions that we call the Moon is about 3,475 kilometers (2,160 miles) in diameter, which is about one-fourth the diameter of Earth. Because of that reduced size, it has about one-sixth the gravity of Earth, meaning that you could definitely dunk a basketball on the Moon, no matter your height or hops.

Despite the fact that it's 848,000 kilometers (527,000 miles) away, the Moon still keeps in touch. It hasn't forgotten about that special day 4.5 billion years ago, so it's maintained contact with us in several ways. For example, the Moon's gravity, though it may be weak compared with Earth's, pulls on our oceans from afar and causes the shifts in our tides. Its gravitational pull also helps to keep us from burning and/or freezing to death. Earth wobbles a little bit on its axis, and the Moon's gravity keeps that wobble in check, which has given us a relatively stable climate.

By the way, in case you haven't heard, we put people on that fucker. In 1969 we sent three guys up to the Moon so two of them could hop around, ride go-carts, drop legendary quotes, and collect Moon rocks.

Unfortunately, we've got a lot of work to do before we set up permanent shop there. Without even touching the subject of how limited our space-traveling capacities are (sigh), the Moon has no atmosphere. Temperatures swing wildly over there. It gets extremely toasty when sunlight hits it and extremely frosty when the sunlight goes away. Not only that, but the whole lack of atmosphere thing makes it difficult to grow anything. If we can't grow anything, we can't eat.

But don't lose hope. Scientists recently found water—the foundation of life—on the Moon, which could make it much easier to start our lunar civilization when the time is right.

THE MOON ○ NASA

THE APOLLO TURD

transcript.

HERE WE HAVE A SAMPLE OF PURE GOLD FROM THE HISTORY OF HUMAN SPACEFLIGHT.

The year was 1969, and the Apollo 10 crew was on its way to the Moon. Their mission was to orbit the Moon and test all systems in preparation for the Apollo 11 mission that would actually land on the Moon two months later.

We can only imagine how maddening it must have been to get so close to the Moon and know that the next guys would be the ones to actually walk on it. But the Apollo 10 astronauts were dutiful and noble, and they didn't let on if they were pissed off about missing the glorious opportunity that Neil Armstrong and Buzz Aldrin would be getting.

How do we know that they didn't bitch about running the lunar landing's dress rehearsal? They were recorded. Every word spoken between astronauts on the eight-day mission was recorded and transcribed by NASA, and after many years of confidentiality, this transcript was eventually released.

On day four, the astronauts are having a serious conversation about a potentially damaged fuel pump, until . . .

Page 414   CONFIDENTIAL   Day 6

Time	Speaker	Dialogue
05 13 26 50	CMP	Oh, yes.
05 13 26 55	CDR	I don't know whether that will reach with that end.
05 13 28 06	CDR	What about this bear? You think you ... up and down?
05 13 28 07	CMP	What?
05 13 28 10	CDR	This one here.
05 13 28 13	CMP	It probably will, unless we tie it up.
05 13 28 16	CDR	Here, you can - can you double it back on the -
05 13 28 22	CMP	If I had a strap, I'd strap it to this here thing, right here.
05 13 28 24	CDR	You don't want anything coming down and whopping you during that burn.
05 13 28 28	CMP	No. Here, let's get this - let me get - Well, I don't see any more straps.
05 13 28 42	CDR	Here's somebody's ear -
05 13 28 46	CMP	I thought I shit the ...
05 13 29 02	CMP	Yes.
05 13 29 30	LMP	Yes, Tom; I don't know. I'm not averse to making - to getting - you know, once you get out of lunar orbit, you can do a lot of things. You can power down, you can do a lot of things. ... another fuel cell, we're going to be sucking a hind tit. And what's happening is - -
05 13 29 44	CDR	Oh - Who did it?
05 13 29 46	CMP	Who did what?
05 13 29 47	LMP	What?
05 13 29 49	CDR	Who did it? (Laughter)
05 13 29 51	LMP	Where did that come from?
05 13 29 52	CDR	Give me a napkin quick. There's a turd floating through the air.

CONFIDENTIAL

Day 6 ~~CONFIDENTIAL~~ Page 415

05 13 29 55	CMP		I didn't do it. It ain't one of mine.
05 13 29 57	LMP		I don't think it's one of mine.
05 13 29 59	CDR		Mine was a little more sticky than that. Throw that away.
05 13 30 06	CMP		God almighty.
05 13 30 08	SC		(Laughter)
05 13 30 10	CDR		What do you see?
05 13 30 12	CMP		Nothing, that's enough for me.
05 13 30 16	LMP		Yes.
05 13 30 18	CMP		Nice going there.
05 13 30 20	LMP		No more turds are going to fit in there.
05 13 30 23	CDR		Is that waste compartment full?
05 13 30 26	CMP		No, hell; there's nothing in there.
05 13 30 28	LMP		It goes all the way down to the -
05 13 30 30	SC		(Laughter)
05 13 30 32	LMP		Hell, when I got in there, I had to stick my hand in there and ... - He put it in the bag, didn't he? You guys been trying to stick it through there with your fingers?
05 13 30 40	SC		(Laughter)
05 13 30 44	CDR		Okay. Soon as we get contact - -
05 13 30 46	LMP		Tom, what bothers me about this pump action is ... - -
05 13 30 50	CDR		There it goes. Okay. Let's - -
05 13 30 53	LMP		- - is this damn thing - I think it's the pump - is cycling on and off. It's not like it's set, set within limits. The son of a bitch is going on and off, on and off. How long is that going to last? It's like I'm watching the voltage going or something.

~~CONFIDENTIAL~~

ALEXANDRA ENGLISH

TYCHO BRAHE

A MAGNIFICENT BASTARD

Tycho Brahe lived in Denmark in the 1500s and was born into the kind of self-perpetuating wealth that those times were famous for. He could do whatever the fuck he pleased, and he did astronomy.

But he didn't use a telescope because those weren't really around yet—he just looked up at the stars and noted the position of each one. But he was a huge perfectionist, and the measurements he made were way more precise than anyone else's at the time.

He made important observations of things like supernovae, and he actually coined the term *nova*. But he's most famous for the impact his notes had after his death. Johannes Kepler was Brahe's assistant and used his observational data to figure out the laws of planetary motion, a huge leap forward in understanding what the fuck is going on in space.

But Brahe isn't badass just because he made important contributions to science. He is badass because of everything else about him. By most accounts, he was an arrogant lush who liked astronomy almost as much as he liked drinking and fighting. A few tidbits:

Brahe had a pet elk that he'd party with. The elk died after drinking too much beer and falling down a set of stairs.

Brahe had a brass prosthetic nose, because he lost his original nose in a fight. That happened after an argument with another Danish nobleman over the legitimacy of a mathematical formula. Both were passionate drunkard nerds, and neither could prove they were right . . . so in classic 1500s style, they settled it in a duel. They went at it with swords in the dark, and Brahe lost his nose.

Even Brahe's death is deliciously badass. He was at a dinner party, eating and drinking excessively, and needed badly to pee. He didn't want to be rude by getting up before his host, so he held it in. But when he finally got a chance to relieve himself, he'd fucked up his bladder by holding it too long and couldn't go. His bladder eventually burst, poisoning his blood and killing him.

His life was so sordid that since his death, he's been exhumed twice to check whether he was poisoned. One main suspect was Kepler. While he was alive, Brahe always refused to give Kepler his notes, because he was jealous of the little upstart. So it wasn't until Brahe died from partying too hard that Kepler was able to get his hands on the notes and essentially kickstart the scientific revolution.

The other person of interest was the Danish king, who suspected Brahe of sleeping with his mother—and potentially even being his own father.

People like Tycho Brahe really make you wonder whether you're living your life to the fullest.

CHECK THIS SHIT OUT

The Moon orbits the Earth on the plane of the equator, so from the northern hemisphere of the Earth everyone's looking "down" at the Moon and from the southern hemisphere, everyone's looking "up." This means the Moon looks upside down if you travel to the other side of the planet.

GORDON AULD

THE MARS ROVERS

LANDING ON MARS IS ONE OF THE BEST WAYS FOR A COUNTRY TO PROVE THAT IT'S A BOSS DOG.

It's one of the hardest things to do in space: if you can land on Mars, it shows that you've got smart people and lots of money. As a species, we have littered Mars with the corpses of many robots that have been smashed to bits in horrific crash landings.

What makes landing on Mars so difficult is the Martian atmosphere. It's thick enough to create friction for anything passing through it, burning the shit out of anything trying to land. But it's not thick enough to slow down a descending robot much . . . so by the time that burnt-up spacecraft gets to the ground, it's still moving at breakneck speed. It takes a fair amount of genius to figure out how to land a robot on Mars without burning it to ashes or smashing it to smithereens.

This is why you've got to have all the top-drawer geniuses working for your space program if you're going to pull it off. And as it turns out, it helps if those geniuses have the imaginations of ten-year-olds.

When NASA's now-defunct Opportunity rover landed on Mars in 2004, it first tore through the thin atmosphere with a heat-resistant shield. Once it got close enough to the planet's surface, parachutes deployed to slow it down. Finally, about twelve meters (forty feet) above the surface, a bunch of airbags made of super-tough material rapidly inflated all around it, and when it hit the ground, it bounced for about a quarter of a kilometer (820 feet) until it had lost all its speed. You can just imagine the faces of the other engineers when one of them suggested covering the 400-million-dollar robot in balloons: "Just let it bounce around a bit, it'll be fine!" But they weren't wrong! Opportunity landed safely and operated on Mars for more than fourteen years.

In 2012, when NASA landed the even bigger Curiosity rover on Mars, the kid dreaming up the insane landing got a bit more sophisticated. This time, once the parachutes had slowed down the capsule enough, the "Skycrane" was deployed. Retro rockets were fired out of the upper part of this contraption like an enormous jetpack, slowing the lander down even more until it was hovering above the surface. Then, cables extended from the bottom to lower the rover down and gently plop it on the surface. It was based on the technology we use in helicopters to lift people out of the ocean and that kind of thing. But instead, they used it to put a van-sized scientist robot on another planet.

The coolest thing about Mars landings is that they all have to be preprogrammed. When something is on Mars, it takes about seven minutes to send any kind of information to or from it. So there's no way to know in time if a tiny course correction is needed or if you're about to land on a massive boulder or if part of the crane isn't working as expected. Mission engineers have to map out the surface using orbiting satellites, figure out the safest places to try to land, and then set the coordinates and the timing of the parachutes, retro rockets, balloon inflation, and everything else way in advance. It's like creating a massive and complex domino setup, knocking over the first one and then having to just wait and see if they all successfully fall down.

But if it does work, and the robot gets safely to the Martian ground, it's so worth the effort. The landers and rovers that have explored Mars so far have taught us a ton about the planet next door, proving that it used to have water on the surface and could once have supported life. And most importantly, they've proven for all to see that their makers are badass geniuses and that no idea is too crazy to try.

ROCKS AT PAHRUMP HILLS, MARS ○ NASA/JPL/UNIVERSITY OF ARIZONA

THE VOYAGER MISSIONS

Voyagers 1 and 2 are spacecraft that were launched in 1977, sent to explore the outer planets of our solar system. And they are, without a doubt, two of the most glorious, badass, and beautiful things humans have ever done in space.

KAREENA ZEREFOS

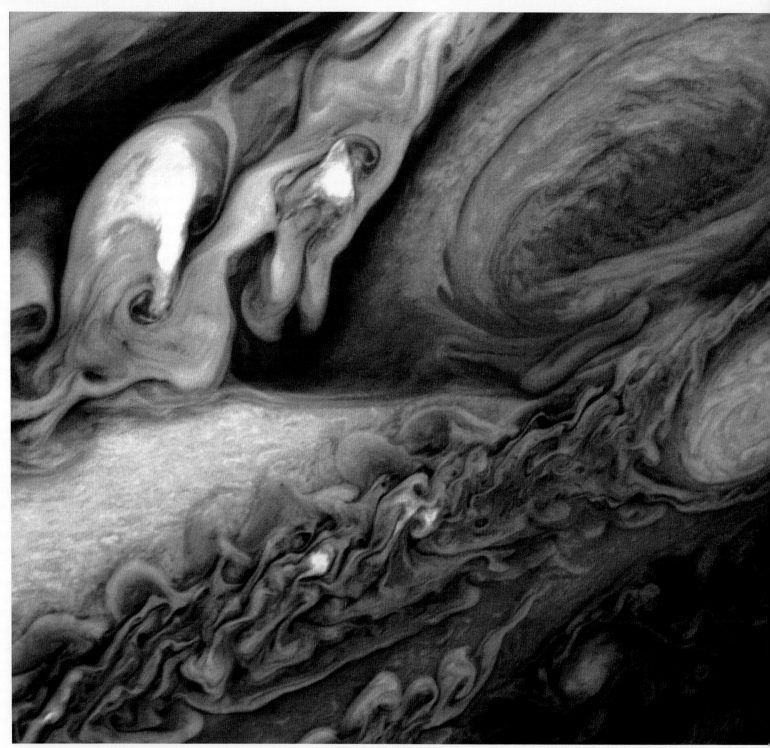

JUPITER'S ATMOSPHERE CAPTURED BY VOYAGER 1 ○ NASA/JPL-CALTECH

One element that is so impressive is the incredible amount of bang they got for their buck. Usually when you want to learn something about a planet, you send a spacecraft there; it goes into orbit, studies that one planet, and maybe learns about its moons. But that's it.

In the case of the Voyager program, they sent two spacecraft off to visit four planets—including Neptune and Uranus, which we'd never seen up close before. They also checked out a bunch of moons and then just kept on going out into interstellar space.

The reason they were able to do this is that the planets aligned in a totally perfect way. That's the kind of thing your astrologer usually talks about, but in space exploration, planetary alignment really just means that good timing can create an enormous shortcut between planets. Planets orbit the Sun at different speeds, and so sometimes they are on opposite sides of the Sun from one another, while other times they're right near each other. It's a bit like the second hand on a clock moving quickly around the numbers, while the minute hand moves more slowly. They still pass each other once every minute. And once every hour, you get the hour, minute, and second hands all lined up.

When the Voyager missions set off, it was a bit like that. The opportunity came up to launch at a time when Jupiter, Saturn, Neptune, and Uranus were all going to be lined up on the same side of the Sun. All we had to do was launch one spacecraft—but hell, we made it two—and swing past them all in one shot.

But it was a bit more complex—and awesome—than that. Every time the Voyagers passed a planet, they'd swing around it, drawn in by its gravity but not enough to be pulled into orbit. They'd use this gravitational pull to accelerate and shoot off in a slightly different direction. So the course each spacecraft took wasn't totally straight. Nonetheless, each slingshot (the actual scientific term) around a planet gave them the acceleration they needed to get as far as they have.

And boy, did they ever get far. The two sister spacecraft opened our eyes to the outer solar system, showing us places we'd only ever seen as dots in a telescope.

They sent back images and data of worlds we'd never even dreamed of, like two of Jupiter's moons: Io, covered in volcanoes, and Europa, covered in scarred ice. They also gave us our first close-up views of Uranus and Neptune, weirdly smooth planets that, even up close, retain a lot of their mysterious allure. The two Voyager spacecraft zipped on by these planets and moons and showed us just how crazy our little corner of the universe is. Then they kept on flying off into the cosmos.

The Voyager 1 spacecraft, which took a shorter route past Jupiter and Saturn, skipping Uranus and Neptune to just haul ass out of the solar system, is now the farthest manmade object from Earth. It has overtaken everything sent out before it, including the Pioneer 10 and 11 spacecraft, which were launched into the outer solar system years earlier. Teams of scientists have continued interpreting data coming from Voyagers 1 and 2 since their launches in 1977 and have determined that both probes have officially left the solar system.

And they're just going to keep going! Both Voyagers are hurtling through space at over 17 kilometers (10.6 miles) per second right now, which is a nutty speed to try to wrap your head around. And there's nothing to slow them down or stop them. In about a million years, either spacecraft might smash into a planet, get sucked into a black hole, or pass too close to a star and disintegrate. The chances of any of that happening, though, given the huge percentage of the universe that is occupied by absolutely nothing, are incredibly slim. But given enough time—and let us tell you, the universe has a lot of time on its hands—even something this unlikely is bound to occur.

One of the more interesting possible futures for these spacecraft is that they may someday be intercepted by an intelligent alien species. The scientists working on the Voyager program thought about this possibility and put a message on each spacecraft for those hypothetical aliens. Each Voyager has on its body a 12-inch golden phonograph record—this was the '70s, remember—with pictures and sounds of Earth written into it. On its cover it shows instructions for playing the record and for finding the location of our planet, written in mathematical language that they figure any intelligent species should be able to decipher.

The Voyager golden records are a sort of time capsule of Earth, life, and human culture.

Each contains images of animals, sounds of waves crashing, recordings of songs, photos of people of all cultures, and all kinds of other things to try to create a snapshot of what is going on here on this planet—what makes Earth, Earth. You can actually find everything they included on the record online, and it's a sweet collection.

Not surprisingly, Carl Sagan, a magnificent and beautiful scientist, led the golden record project. He and his team understood the significance of what they were doing by sending this spacecraft off on a course that would take it out of our solar system. By including with it some kind of picture of Earth, its life, and its culture, they were also able to preserve it in some way. So even when the Sun gobbles up the Earth, which it will inevitably do, a human artifact will continue to exist in the cosmos. Even if that message in a bottle is never intercepted, Sagan and his team made the human species, our home, and our culture immortal in the universe.

LIVING OFF-PLANET

Someday in the future, if we don't fuck up too much right now,
humans will be living off Earth in large numbers.
The exact way in which we'll live off-planet could vary.
A future person looking to pick somewhere to settle down
will have some interesting options.

MAYBE YOU LIVE IN AN ORBITAL SPACE COLONY?

Right now, the only way humans live in space is on space stations—big, complicated boxes floating in orbit around the Earth. These are nice because they're so close to home. They can get supplies easily, and if shit hits the fan, the people on them can come home pretty quickly. In future times, we could have bigger, more complicated, more comfortable space stations orbiting the planet, with artificial gravity and freshly grown space-food.

In the 1970s, futuristic space visionary Gerard K. O'Neill proposed nifty space habitats that we now call O'Neill Cylinders. The idea is to fit entire space colonies into huge rotating cylinders. Imagine a tube from inside a bike tire, but big enough to fit a forest inside. Then spin it so that it creates fake gravity on the inside, fill it with air, and build some towns in there. It's a totally not-crazy idea that we might make a reality, if we make it out into space in big enough numbers.

LIZI PRATT

CAL SINADINOVIC

MAYBE YOU LIVE ON A STARSHIP?

These days, we're making lots of progress finding planets around other stars, and many of them are similar enough to Earth that we could potentially live there. Now that we have a bunch of candidate home planets, the big issue is getting to them. To get even just to the closest star, it would take more than four years at the speed of light. Since we can't even come close to that speed, it would probably take us tens of thousands of years to get to another star system.

Plenty of smart people have spent a lot of time dreaming about a future scenario like this and have come up with some sweet designs for the spaceships that would carry generations of people from point A to point B.

Imagine: you board a massive spaceship capable of carrying you and thousands of others for your entire life, live there, fall in love, have some babies in space, and die knowing that only your distant descendants will actually get to live on the planet you got on that spaceship to reach.

Even more crazy: imagine being born on this spaceship in one of the boring generations between the two points. You live your entire life on this spaceship in the frozen wasteland of outer space, hearing stories about where your ancestors came from and where your great-great-great-great-grandchildren will arrive.

MAYBE YOU LIVE ON MARS?

As it is right now, Mars isn't a super-appealing place to live. Well, maybe it's better than a starship. But still, it's freezing cold, you can't breathe the air, nothing grows there, and there's some pretty serious radiation coming at you from the Sun. All of this can be blamed on one thing: Mars doesn't have much of an atmosphere. Our atmosphere is what keeps Earth all nice and warm and wet, and without it, Earth would be a lot like Mars. So people have thought about how we might make Mars a bit more liveable by doing what we call terraforming—that is, making it more Earth-like.

The terraforming ideas being kicked around these days are pretty fucking awesome. One involves capturing comets in the solar system and redirecting them so they smash into Mars and create heat, which would evaporate ice, turning it into water, and build up an atmosphere. Another idea is to send a bunch of specialized bacteria to Mars that can eat its soil and fart out oxygen, eventually building up into an atmosphere we could breathe.

Variations on these kinds of ideas exist for just about all the other planets and moons in our solar system. Maybe we have a floating colony in Jupiter's clouds? Or an underwater colony in the depths of Ganymede's ice-covered oceans? No matter which idea we choose, it's going to be dope.

STORE

MICRO ORGAN ISMS:

AS THE TRUE MASTERS OF SPACE COLONIZATION

MONIXX DESIGN

IF ANY EARTHLING IS GOING TO COLONIZE SPACE, PUT YOUR MONEY ON MICROBES.

WE HUMANS ARE <u>PATHETIC PIECES OF SHIT</u> NEXT TO MICROBES.

On Earth, not only do microbes outnumber us but they also make up a shocking amount of what we consider "us." There are twice as many microbe cells as human cells in your body, and there are in fact so many microbes inside you that half the mass of any given turd you push out is actually microbes.

Take a second to process that before we move on.

When it comes to space, microbes are way more adaptable than us. Experiments have shown that a huge number of microorganism species can straight up live in the vacuum of space, at least for a while. During the Apollo program in the 1960s, astronauts brought back a camera that had been on the Moon for a couple of years and found living bacteria that had survived since before launch. They hadn't sterilized the camera properly before sending it to the Moon, but nobody thought it would matter because we thought conditions on the Moon were totally inhospitable to life.

As it turns out, a lot of microbes do just fine in conditions that we consider deadly. There is a whole category of microorganisms called extremophiles, freaky little creatures that thrive in extreme environments like pools of acid or boiling hot ocean vents. There are even microbes that live in nuclear waste without any oxygen. Basically anywhere we go on Earth, we find microbes that seem right at home.

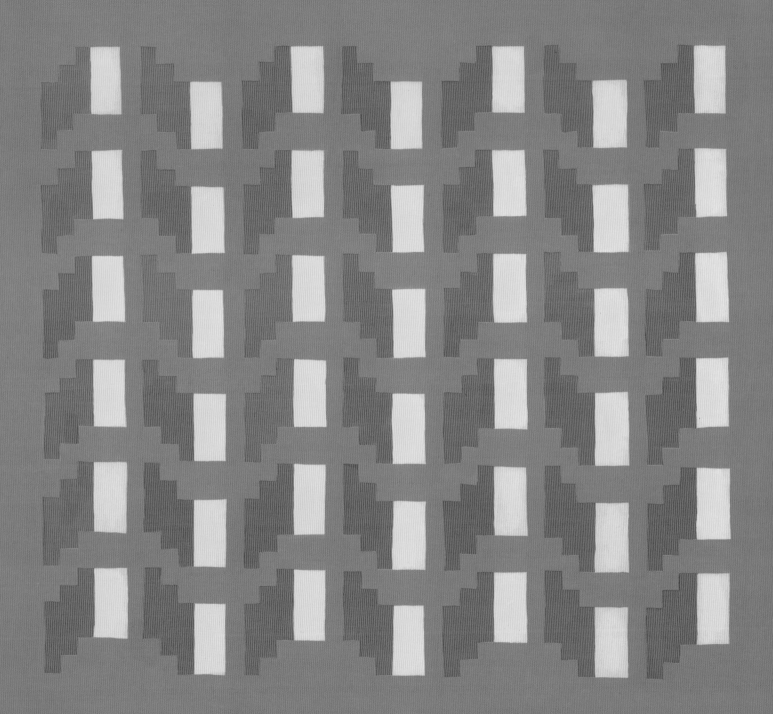

CARL JIORJIO

At this point in the search for life beyond Earth, we're pretty sure we're not going to find the Martians or Venusians (or Grays) that sci-fi has traditionally imagined. What we're looking for now are microbes living on other planets. We think they could exist below the surface of Mars, for example, or in the subsurface oceans of Europa or Enceladus. If there's extraterrestrial life in our solar system, it's probably microscopic.

But there's also a very real chance that instead of finding microbial life on another planet, we're going to put it there. NASA and other space agencies all sterilize their spacecraft to kill off all the microbes, and there is even a Planetary Protection Officer at NASA, whose job is to make sure we aren't contaminating these other worlds. But it's really fucking hard to totally and completely sterilize anything, and we definitely haven't always been this careful with it. We might have unwittingly colonized Mars in the 1970s when we started sending landers there, thus starting the evolutionary path toward future Martians.

So it's not farfetched to think microbes will colonize space instead of us. Microbes truly are the masters of Earth life. They totally dominate us and every other species, despite what our human egos tell us. We might as well acknowledge that they're the rightful masters of the rest of the solar system too.

CARL JIORJIO

EARTH'S MAGNETIC FIELD

ARTIST'S RENDITION OF EARTH'S MAGNETOSPHERE ○ SOURCE NASA/GSFC

Earth's magnetic field is pretty fucking cool.

It's a whirling system of magnetic currents that surrounds the planet, generated by the motion of molten iron in the Earth's outer core. It's nature at its coolest. Liquid metal in the middle of our planet sloshes around as the planet turns, generating what is actually called a fucking dynamo through the power of physics.

This magnetic field does all kinds of cool shit, like guiding your compass needle and making the pretty colors of the auroras. Birds use it to guide their migrations, and scientists have even noticed that some animals prefer to lie down along the lines of the magnetic field—that is, they line their bodies up north–south. But beyond this cute shit, the magnetic field also does the hugely important job of keeping us all alive by protecting our atmosphere from getting ripped off by our other powerhouse friend, the Sun.

The Sun, as we all know, is a giant beast of a fusion reactor that ought to be feared and respected. It does all kinds of insane shit that keeps you alive while also being able to kill you in a fucking millisecond if you find yourself in the wrong place in space. Along with all that lovely sunlight, the Sun creates solar winds, streams of charged particles shooting out in all directions at up to 1,000 kilometers (621 miles) per second. And even puny little particles can do some serious damage at those speeds.

They'd tear our atmosphere right off the planet and throw it into space, killing us all—that is, if it weren't for our good buddy the magnetic field, which deflects the particles.

So the magnetic field acts like a sort of shield, creating a safe bubble around the Earth that we call the magnetosphere.

It's weird to think that we'd all be killed by particles shooting out of the Sun if it weren't for some liquid iron moving around about 3,000 kilometers (or roughly 1,800 miles) below us, but that's the kind of funky shit that happens around here.

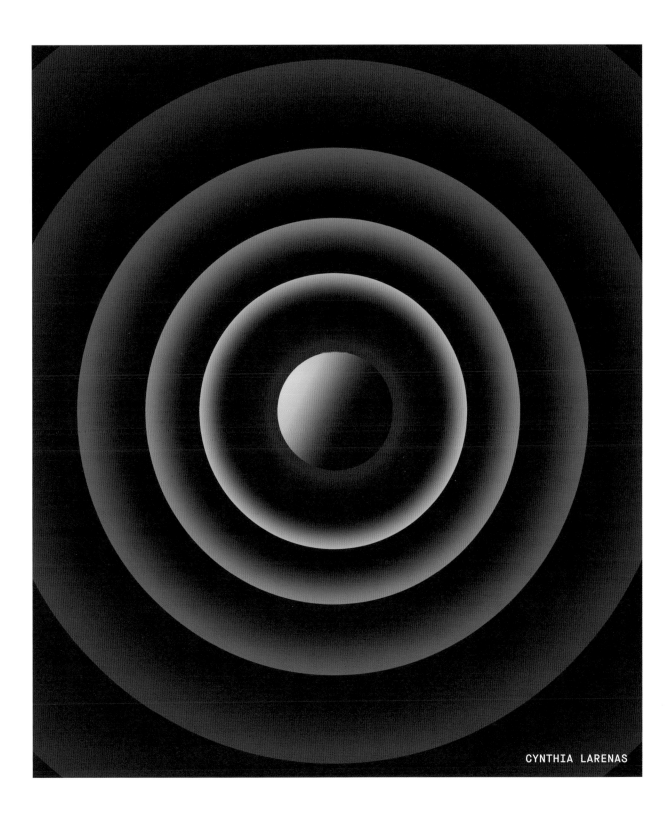

THE SUN

SYNC

AIDA AZIN

MANY ANCIENT CIVILIZATIONS WORSHIPPED THE SUN AS A GOD—AND FOR GOOD REASON. THAT BRIGHT BALL OF SUPERHEATED GAS UP THERE IN THE SKY IS WHAT KEEPS LIFE GOING HERE ON EARTH.

Science's best estimates put the Sun at about 4.6 billion years old, and although it takes up just a small portion of the sky during the day, that motherfucker is big. It's a million miles (1.6 million kilometers) across, and about a million Earths would fit inside it. But as big as it is, it's actually quite average, or even a bit small, compared with other stars around the universe. Astronomers with a knack for cute names have classified it as a yellow dwarf.

But don't let the cute name fool you. It can fry the fuck out of your skin from 150 million kilometers (93 million miles) away, so just imagine what it can do from up close. Its temperature stays at a toasty 5,500°C (nearly 10,000°F) at the surface and then hikes up to 15 million°C (27 million°F) in the core.

How does it get so fucking hot? Nuclear fusion. You see, the Sun is really big and really dense, so it's got a lot of energy. A whole lot of energy. And if you throw a bunch of hydrogen atoms into a really dense place with lots of energy, they flip out and start smashing into each other really hard. When two hydrogen atoms smash into each other hard enough, their nuclei fuse together and they make a helium atom. This smashing and fusing releases a shitload of even more energy, which makes the Sun the explosive inferno of hot gas and plasma that it is.

So the Sun is massive, hot, and does nuclear fusion. What's the big deal? The big deal is that without the Sun, Earth would be a lifeless wasteland. There are a lot of reasons for this, but one of the biggest is this thing called the food chain. Plants get their energy from the Sun, so if there's no Sun, there are no plants. If there are no plants, we animals don't eat, and we die. This means that we owe our very existence to that chaotic sphere of colliding atoms in the sky.

We should appreciate it while it lasts too. Like all living things, stars have life cycles, and this means that one day, our Sun will die. Before it does that, though, it's going to expand into what those astronomers call a red giant. In around 5–7 billion years, it's going to balloon outward in size and engulf our entire planet, scorching the surface and turning our home into a dry, molten rock, void of any life.

But don't lose sleep over it. The odds are that some other cataclysmal quirk of the cosmos will claim humanity before then.

REALLY BIG
&
REALLY DENSE

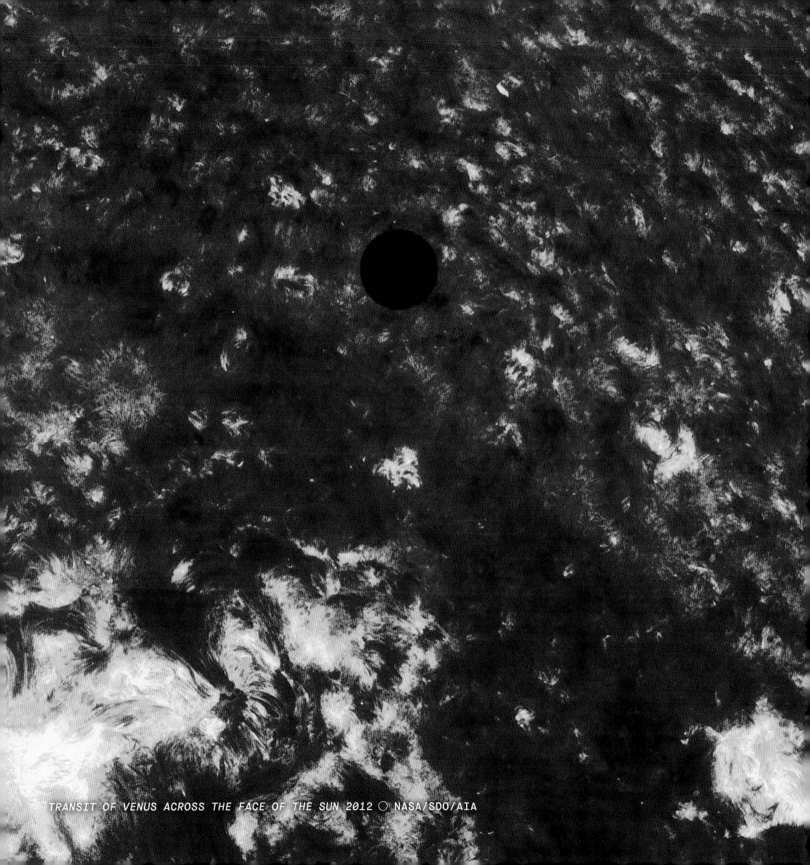

TRANSIT OF VENUS ACROSS THE FACE OF THE SUN 2012 ○ NASA/SDO/AIA

VENUS
AS A LITERAL HELL

Back in the day when we were first using telescopes to look at other planets, we looked at Venus and couldn't see the surface because the atmosphere was so thick.

So we thought, "Oh cool, it's covered in clouds. It's probably like Earth. But it's pretty close to the Sun, so it's probably pretty hot in there. It's probably like a hot jungle swamp or some kind of tropical paradise. Yeah, it's probably all tropical and humid with all kinds of swamp-dweller creatures on it—alligators and shit. Maybe when we send humans there in like ten or fifteen years, we'll set up neat little swampy resorts, and people can take their kids to go look at the Venusian gators."

That was the verbatim official position of our leading scientists. Look it up.

Anyway, it pretty much goes without saying that this was way off. But scientists believed this until the 1960s, at which point they developed radar imaging technology and were actually able to detect what was really going on under Venus's clouds.

And what they saw was pretty much hell.

The surface of Venus isn't just some kind of uncomfortably hot jungle-like place. It's hell.

First of all, the atmosphere is crushingly dense, so when you're on the surface of the planet, the weight of the atmosphere amounts to the same pressure as being about a kilometer (over half a mile) underwater. That's lung-crushing pressure. We've sent spacecraft to study Venus that never even made it to the surface because they were crushed by the atmosphere before they got all the way through. This is some serious shit.

But that's not even the half of it. The average temperature on Venus is about 460°C (860°F). The crazy atmosphere creates a greenhouse effect, trapping the heat of the Sun and just getting hotter and hotter. Scientists think that Venus might have been pretty chill way in the past—maybe even a bit like Earth.

But now it's so damn hot that metals turn to gas at the surface, rise up into the atmosphere, and fall down again as metal snow. When it's not snowing metal, it's raining sulfuric acid. Oh, and hey, the clouds are made of sulfur dioxide, so the whole place smells like farts.

Some lunatics still want to go there, though, and have sketched out plans for habitations that float high in the atmosphere, between layers of stinky clouds. Knock yourselves out, we say.

CHECK THIS SHIT OUT

```
Venus spins backward compared with other planets. And it takes
243 Earth days for the planet to rotate on its axis just once.
But it takes just 224.7 Earth days for it to go once around the
Sun, which means on Venus a day is longer than a year.
```

CAROLINE LEVASSEUR

SATURN, YOU BIG, BEAUTIFUL BASTARD

RICHARD AMEZQUITA

Ugh, Saturn. Stop. You are just too much. You've got everything, Saturn.

You've got swirly clouds.

A STORM AT SATURN'S SOUTH POLE ○ PD/US GOVERNMENT/NASA

You've got a storm shaped like a hexagon on your north pole.

TRUE SATURN ○ SOURCE UNKNOWN

A GLOBAL STORM GIRDLES THE PLANET ○ NASA/JPL-CALTECH/SPACE STATION INSTITUTE

You've got a storm that could hypnotize a god on your south pole.

You've got shitloads of moons and super-pretty rings.

Oh Saturn, you're so beautiful you make me want to cry.

NORTH POLAR HEXAGON ○ NASA/JPL-CALTECH/SPACE SCIENCE INSTITUTE

BIG & SMALL SPACE ROCKS

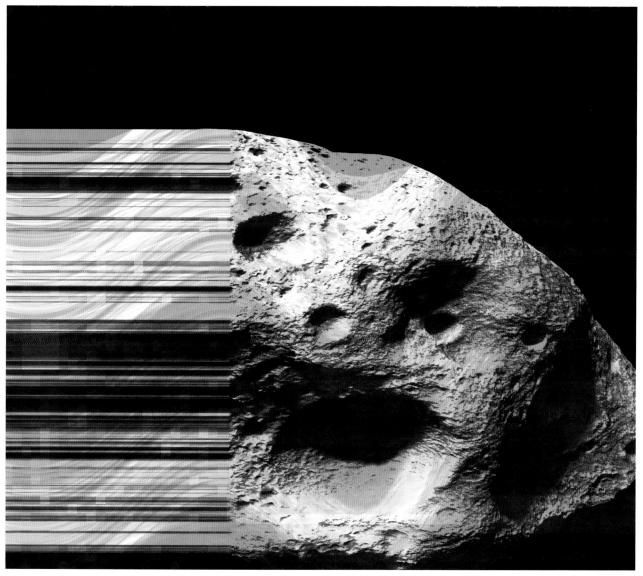

HI$TO

COMETS

A comet is a hunk of ice and dust and frozen gas that floats through space, usually hanging out far away from the Sun (or whatever star it's orbiting). But sometimes a comet gets too close to the Sun and heats up, turning the ice into steam and freeing up frozen gas and dust. All this gas and dust gets blown away from the comet by the Sun's energy, which makes that nice tail we sometimes get to see from Earth.

METEORS

A meteor is a little bit of space rock burning up in the Earth's atmosphere—AKA a shooting star. If the little guy makes it to the ground without getting all burned up, we call it a meteorite. Seeing a shooting star is actually pretty fucking sweet: you're seeing a rock that was forged in the heart of a star and has traveled millions of miles through space, suddenly getting totally obliterated by the air you breathe.

BILLIE JUSTICE THOMSON

ASTEROIDS

An asteroid is a big motherfucking space rock. Most of the asteroids in our solar system hang out between Mars and Jupiter, but some of them break out of this orbit and go cruise around elsewhere. Sometimes they go smashing into planets. An asteroid smashed into the Earth a few million years ago and we all know what happened there.

ASTEROID LUTETIA ○ ESA 2010 MPS FOR OSIRIS TEAM MPS/UPD/LAM/IAA/RSSD/INTA/UPM/DASP/IDA

ASTEROID VESTA ○ NASA/JPL-CALTECH/UCLA/MPS/DLR/IDA

THE PROBLEM

Asteroids are the number one big scare for Earth because there's a shitload of them in our neck of the woods and they have the power to totally wipe out life on a massive scale. We've identified and are tracking a lot of them, but there are a lot more of them, most of which are small (on the cosmic scale) and dark, so they're hard to spot. It's not uncommon for us to spot an asteroid only a few days or weeks before it passes scarily close to us. Some even pass between the Earth and the Moon, and that is fucking close in space terms.

If we found an asteroid that was going to hit the Earth, even a couple of years before impact, the chances are very slim that we'd be able to do anything to stop it. We don't have technology to deflect asteroids, and we can't just go blowing them up *Armageddon*–style because that would probably only create a whole bunch of smaller chunks of rock that would still hit the Earth. It's definitely technically possible to deflect asteroids, but no government has invested in developing those technologies yet, and they take a long while to figure out.

Luckily for us, there are two good guys who can help save us from asteroid strikes: Jupiter and billionaires.

THE SOLUTION

Jupiter is so massive that it sucks up a lot of the junk flying around in space. Space rocks big and small are attracted by its enormous gravity and get drawn into its mysterious, creamy orangeness. It's considered the guardian of the solar system for this reason. Not only is it comforting to know that Jupiter is out there sucking up would-be destroyer-of-worlds asteroids, but it's also really fucking cool to see what it looks like when this happens. The best example came in 2009 when some unidentified space rock about half a kilometer (1,640 feet) across smashed into Jupiter and made a hole. That gnarly hole is inconceivably big—about 190 square kilometers (118 square miles) in area. And true, it's easier to make a hole in gas than in rock, but it is still mind-boggling that a relatively little rock tore a gash that big in a planet.

Now for the other potential protector of the Earth: billionaires. No government so far has stepped up to actually deal with the threat of asteroids hitting the Earth and killing us all, but a lot of entrepreneurs are inadvertently doing so. Their intention is to develop technology to mine asteroids, which are full of stuff that's rare on Earth. They're full of gold, iridium, silver, osmium, palladium, platinum, rhenium, rhodium, ruthenium, and tungsten—shit you know is rare or that you've never heard of because it's so rare. And rare means valuable.

So by looking to make big bucks mining asteroids, these businesspeople are developing technology to wrangle those space rocks and move them around. Which is exactly what you need to do to an asteroid that's on a collision course with Earth.

So cross your fingers and invest your savings in asteroid mining.

SYNC

PLUTO

"BOO HOO," CRIED COUNTLESS LITTLE KIDDIES WHEN PLUTO GOT DEMOTED FROM PLANET STATUS.

Neil deGrasse Tyson, one famous astrophysicist involved in the demotion, even got hate mail from kids demanding that he reinstate Pluto among the planets. And there are still people out there fighting tooth and nail to make Pluto a planetary bro once again.

To these people we say, "Chill the fuck out."

Backstory: ever since it was spotted in 1930, Pluto was considered the ninth planet. It was small, it was cute, and there was even a Disney dog with the same name. No big surprise that people loved it. But in the 1990s and 2000s, scientists found a bunch more teeny guys around Pluto's size. Apparently there's a lot going on out there, and as we get better at looking for planets, we keep finding them.

So rather than expanding the list to include newly found plutoids like Charon and Eris, the space bosses of the world (the International Astronautical Union) decided to make the definition of a planet a bit stricter. To qualify as a planet, a space thing had to meet three criteria:

1. Orbit the Sun.
2. Be round (which happens when you're big enough that your own gravity pulls every part of you equally into the center).
3. Be the boss dog of your orbit.

And Pluto, sadly, didn't meet that third condition.

A planet, by this new definition, had to be big enough to have "cleared the neighborhood of its orbit," meaning that it either bashed everything else near it out of the way or absorbed it all into its own mass. And Pluto hasn't done this.

Therefore, Pluto was kicked out of the planet party, and a surprising number of people freaked out.

To us, this freak-out reaction is silly. It's just a matter of terminology. Pluto is now a "dwarf planet" instead of a full "planet," but this doesn't mean that it's less important or less scientifically interesting. Kids in school won't have to memorize its name, but anybody who actually gives any fucks about space will still learn about it.

In fact, by reclassifying Pluto as a dwarf planet, we've drawn attention to the other dwarf planets. There's a bunch of them out there and they're totally fascinating little worlds (with cool names like Makemake).

So instead of being the shrimp in the family of planets, Pluto is the most famous, most beloved, biggest, bossiest member of its own family of dwarf planets.

So please, shed no tears for Pluto.

CONOR DONALDSON

KATE KURUCZ

DANILO BRANDÃO

Jupiter is a massive beast of a planet that puts the rest of the solar system to shame.

Here are a few fun facts:

JUPITER IS FUCKING ENORMOUS!
It's more than twice as massive as all the other planets combined. If it were about sixteen times bigger, it would be so massive that it would start fusing the atoms in its core and turn into a dwarf star.

JUPITER IS FUCKING HOT!
It's not big enough to fuse atoms, but it sure as hell is crushing them, which generates a shitload of heat. It makes more of its own heat than it gets from the Sun.

THIS HUGE MOTHERFUCKER IS MADE OF GAS!
If you tried to land on Jupiter, you'd go through cloud layers like you would on Earth. But instead of passing some clouds and then landing on a solid surface, you'd just keep going through layer after layer of cloud. And as you descended, the clouds would get thicker and thicker until eventually you'd be in a liquid. There'd be no surface to that liquid, though—it wouldn't be like falling through the air into a lake on Earth. You wouldn't be able to pinpoint exactly when you went from gas to liquid; you'd just be thinking, "Wow, this is some thick steam," and then eventually realize that you were actually fully in liquid. At this point, though, you'd have to be in a supremely strong spaceship to avoid being crushed by the weight of all the layers of gas on top of you.

Similar to the bizarre gas-liquid continuum, there's no clear distinction between where Jupiter ends and space starts. The gas planet just thins out until it's barely there anymore, and then eventually isn't there anymore. Jupiter is basically a continuum between nothingness and intense, crushing density.

JUPITER IS MYSTERIOUS!
We don't even know (yet) exactly what's in the center of the planet.

JUPITER IS A WORLD OF STORMS!
The reason Jupiter is so pretty is that it's covered in insane storms. All the different bands are massive storms moving in different directions. The "Great Red Spot" (a very official scientific name) is a single storm more than twice as big as the Earth and has winds of more than 600 kilometers (373 miles) per hour, which is a fucking furious wind.

Jupiter is about ten times as wide as the Earth but does a complete rotation in only ten hours. This means the gases in the outer parts of the planet are moving around the center of the planet at crazy speeds. This kicks up storms unlike anything we see on Earth.

MOONS OF JUPITER

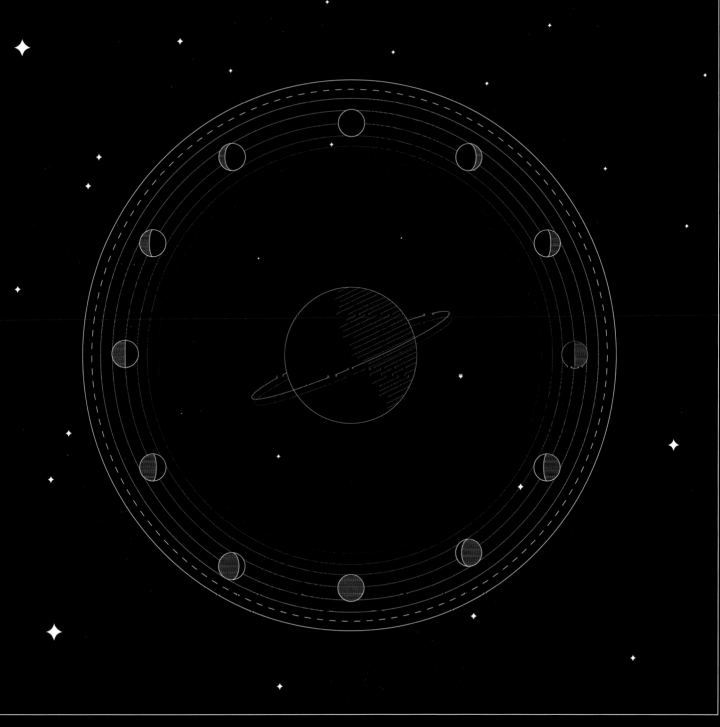

IO ○ NASA/JPL/UNIVERSITY OF ARIZONA

CHECK THIS SHIT OUT

Dark patches on Earth's moon are lava flows from old volcanoes.

EUROPA ○ NASA/JPL-CALTECH/SETI INSTITUTE

JUPITER'S MOONS HAVE ALWAYS BEEN HOT SHIT IN THE WORLD OF SCIENCE.

Galileo (1564–1642) was among the first—if not the first—to spot the moons of Jupiter, and from this observation he reasoned that Earth is probably orbiting the Sun and is not the center of the universe. This pissed off the Pope, and Galileo spent the rest of his life under house arrest. He eventually took it all back, according to the official record—he was practically forced to—but spent the rest of his life sitting around, having people over, and hinting at Earth's orbit around the Sun by quietly saying gangster rebel shit like, "And yet it moves."

We've known about Jupiter's moons for a long time, yet we keep finding more. So far we've counted around seventy moons, most of which are puny little asteroids that just got sucked into orbit. The four heavy hitters are Io, Europa, Ganymede, and Callisto, and it wasn't until the Pioneer and Voyager spacecraft passed them in the 1970s that we actually got a good look at them. And these are some pretty fucking dope moons. Our favorites are Io and Europa.

IO

Pretty much hell, Io is relatively small and orbits super close to Jupiter. And close to Jupiter is a fucked-up place to be, since the planet emits crazy amounts of radiation and has enormous gravity. Since Io is stuck between this monster planet and a few other big moons, it gets tugged on by all of their gravitational pulls, and all these forces acting on it makes it extremely hot and incredibly geologically active. Io, which is about a quarter of the size of Earth, has mountains taller than Mount Everest and about four hundred volcanoes, some of which spew magma hundreds of miles above the surface. We don't see anything like this on Earth, and we'd be fucking terrified if we did.

EUROPA

A bit farther out from Io's ball of lava is Europa, a mysterious world completely covered in ice. These two moons may be neighbors, but they could hardly be less alike. Europa's icy crust is a bit like Earth's rocky crust, except it's way more badass. It's about thirty kilometers (or roughly 19 miles) thick and as hard as granite because of the super-low temperatures caused by being so far from the Sun. On the surface of Europa, temperatures get as low as -220°C (-364°F). But like Io, Europa is being pulled on like crazy by Jupiter and the other moons. This causes friction that warms up the interior, creating a layer of liquid water that is probably around 100 kilometers (over 60 miles) deep and has twice as much volume as the Earth's oceans.

Scientists are nuts for Europa, because it's one of the most promising places in the solar system for finding life. The rule of thumb we use in our quest for extraterrestrial life is that we look for liquid water first, because all life as we know it depends on that. And Europa's got plenty! Plus it's conceivable that life under the ice crust would be protected from the crazy radiation Jupiter's spitting out.

So we're hopeful, although getting a spacecraft to Europa that could drill through miles and miles of ice to check out that ocean is fucking difficult, and therefore expensive. Recently, though, scientists spotted huge plumes of water being shot out into space from Europa. Think Old Faithful, Yellowstone National Park's biggest regular geyser, but shooting water 200 kilometers (over 120 miles) high. NASA is now planning to send a spacecraft to orbit Europa and pass through some of these plumes to see what kind of shit it picks up. Maybe some little microbe buddies. Maybe some space fish. Who knows?

HEAT DEATH

or

BIG FREEZE

A deep dive with an ordinary dude

with guest writer
Garrett Johnson

An optimistic view of the future sees humans getting over our shit, putting an end to war and pollution, and expanding into the universe in the spirit of peace and knowledge. That is a pretty fucking rosy view, but let's imagine it happens.

Even in that case, eventually the universe is going to snuff out our special little candle, and all candles for that matter. No matter how great we become at harnessing the energy of the cosmos, that energy will run out.

Here to explain that bleak inevitability is Garrett Johnson. Garrett isn't a scientist, he's just an ordinary dude. He enjoys pizza pops and making jokes, and he digs space. He reads up on what interests him, and now here he is to teach you all about it.

"Heat Death" and "The Big Freeze" might sound like Arnold Schwarzenegger movie titles, but they're actually two names for the most plausible end of our universe. The fate of our universe is still uncertain, and all the predicted doomsdays are just hypothetical. But so far, the evidence supports one particular outcome. So buckle up, kids, because here's what's going to happen to our precious universe.

One day, the Sun is going to die. It's going to run out of stuff to burn, and with a cough and a wheeze, it's going to die. The stuff it was made out of will just sort of scatter away and eventually get sucked up by a black hole—probably the black hole at the center of the Milky Way galaxy. Eventually all the stars in the universe will die, just like everyone you know, and there will just be a giant black hole where every galaxy was.

And then the black holes will die.

For a long time it was thought that nothing could escape a black hole. But then Stephen Hawking came along and theorized that a very tiny amount of energy might be coming from black holes as radiation. As the black holes emit this "Hawking radiation," they lose mass. After a staggeringly long time (potentially up to 100,000,000,000,000,000,000,000,000,000,000,000,000,000,000,

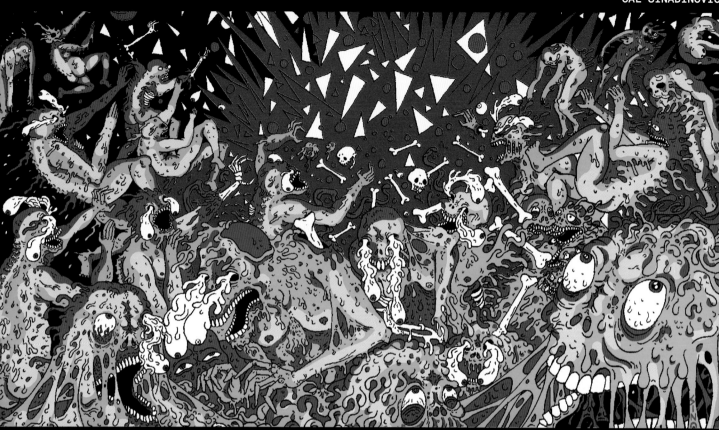

000,000,000,000,000,000,000,000,000, 000,000,000,000,000,000,000 years), a galactic black hole will have leaked out enough radiation that it would actually decay into nothing.

Once all these black holes have died this slow, boring death, the universe will enter a "dark era" and all that's left will be a bunch of extremely low-energy particles floating around and occasionally bumping into an atom or two that somehow escaped the black hole fate. Eventually even these bumps will stop happening, and then, forever more, nothing will happen.

This state is the end result of the universal progression toward entropy—a state of evenly dispersed chaos in which things reach equilibrium. This is why your coffee gets cold and your room gets messy. You have to keep adding energy to maintain that difference in temperature or to stack piles of books. And eventually, you, like the universe, are gonna run out of energy.

CHECK THIS SHIT OUT

The universe has continued expanding ever since the big bang got things going. It's getting bigger and bigger all the time, and it's actually speeding up. Which is a pretty wacky thing to think about.

The universe is, by definition, everything that exists including empty space. So if it's expanding, what's it expanding into? What was out there beyond the edge of the universe before the universe got big enough to occupy that space? I don't have answers for you!

Index of Artists

Aida Azin
— 130

Alex Gvojic
— 52

Alexandra English
— 73, 94, 109

Alice Dolling
— 85

Andrea Hsieh
— 45, 62

Arik Roper
— 30

Billie Justice Thomson
— 142, 143

Brett Randall
— 8, 66

Cal Sinadinovic
— 57, 114, 163

Carl Jiorjio
— 122, 124

Caroline Levasseur
— 8, 135

Chrissie Abbott
— 22, 54

Conor Donaldson
— 96, 149

Cynthia Larenas
— 127, 155, 164

Danilo Brandão
— 152

Duel
— 43

Gordon Auld
— 77, 80, 100, 110

Haha
— 75

Hi$to
— 141

James Marshall
— 15

Kab 101
— 21

Kareena Zerefos
– 105

Kate Kurucz
– 151

Lauren Bradley
– 87

Lee McConnell
– 169

Lizi Pratt
– 113, 118

Marina Zumi
– 38

Matt Lyon
– 27

Matthew Volz
– 70

Mike Makatron
– 28

Monixx Design
– 10, 121

Neecho
– 67

Ngaio Parr
– 166

Nial Gayle
– 5

Ria McIlwraith
– 34, 62

Richard Amezquita
– 84, 137

Rowan Dodds
– 158

Sam Songailo
– 51

Sophie Filomena
– 18

Store
– 117

Sync
– 129, 146

Trevor Baird
– 58, 99

Wildergrim
– 160

LEE MCCONNELL

Kate Howells is a science enthusiast and evangelist whose mission is to get people to feel some awe and wonder about the reality around them. In her day job, she works for an international NGO, The Planetary Society, where she does community engagement and writes about planetary science and exploration. Kate didn't grow up a space nerd but became one in her twenties, proving it's never too late to get obsessed with something. She lives in Guelph, Canada. *Space Is Cool as Fuck* is Kate's first book.